Lecture Notes in Networks and Systems

Volume 48

Series editor

Janusz Kacprzyk, Polish Academy of Sciences, Warsaw, Poland
e-mail: kacprzyk@ibspan.waw.pl

The series "Lecture Notes in Networks and Systems" publishes the latest developments in Networks and Systems—quickly, informally and with high quality. Original research reported in proceedings and post-proceedings represents the core of LNNS.

Volumes published in LNNS embrace all aspects and subfields of, as well as new challenges in, Networks and Systems.

The series contains proceedings and edited volumes in systems and networks, spanning the areas of Cyber-Physical Systems, Autonomous Systems, Sensor Networks, Control Systems, Energy Systems, Automotive Systems, Biological Systems, Vehicular Networking and Connected Vehicles, Aerospace Systems, Automation, Manufacturing, Smart Grids, Nonlinear Systems, Power Systems, Robotics, Social Systems, Economic Systems and other. Of particular value to both the contributors and the readership are the short publication timeframe and the world-wide distribution and exposure which enable both a wide and rapid dissemination of research output.

The series covers the theory, applications, and perspectives on the state of the art and future developments relevant to systems and networks, decision making, control, complex processes and related areas, as embedded in the fields of interdisciplinary and applied sciences, engineering, computer science, physics, economics, social, and life sciences, as well as the paradigms and methodologies behind them.

More information about this series at http://www.springer.com/series/15179

Gheorghe I. Gheorghe
Editor

Proceedings of the International Conference of Mechatronics and Cyber-MixMechatronics – 2018

Springer

Editor
Gheorghe I. Gheorghe
National Institute of Research and
Development in Mechatronics and
Measurement Technique, INCDMTM
Bucharest, Romania

ISSN 2367-3370 ISSN 2367-3389 (electronic)
Lecture Notes in Networks and Systems
ISBN 978-3-319-96357-0 ISBN 978-3-319-96358-7 (eBook)
https://doi.org/10.1007/978-3-319-96358-7

Library of Congress Control Number: 2018948220

This Springer imprint is published by the registered company Springer Nature Switzerland AG
The registered company address is: Gewerbestrasse 11, 6330 Cham, Switzerland

Preface

The 2nd International Conference of Mechatronics and Cyber-MixMechatronics/ICOMECYME was held in Bucharest (Romania), on 6–7 September 2018. This conference is envisioned as a forum and an opportunity to researchers, engineers, professors, PhD students and graduate students as well as business representatives from all over the world to present their research results and development activities. It comes as a consequence of the expansion of the field of mechatronics, which has come to step into the world of newer trans-disciplinary fields of adaptronics, integronics and cyber-mixmechatronics.

Originally entitled "International Conference on Innovations, Recent Trends and Challenges in Mechatronics, Mechanical Engineering and New High-Tech Products Development" (MECAHITECH), the conference was held for the first time in 2009.

With a new name and derived from an event that deeply penetrated into the academic community, gathering specialists from all over the world, including North America, South America and Asia, the conference facilitates reflection on the current state of the addressed fields and discussions about potential future directions for research.

This proceedings volume provides the research community with well-edited, high-quality papers. For the purpose of this book, 21 papers, authored by teams of researchers from universities and research institutes, were selected for inclusion.

The papers presented at the conference will be a ramp for PhD, PhD students, researchers and engineers, who will have to be prepared to crosstraditional boundaries in order to accommodate the new technologies. This volume will thus be a valuable addition to the literature, and it will examine the intersection between mechatronics, cyber-mechatronics and cyber-mixmechatronics, as well as other related disciplines, and it will assess the implications for industry throughout the world.

I am particularly grateful to the authors for their contributions and all the participating experts for their valuable advice. Furthermore, I thank the staff for their cooperation and support, and especially, all members of the International

Programme Committee and the Organizing Committee for their work in preparing and organizing the conference.

On behalf of the Organizing Committee, I would like to thank Springer for its professional assistance and particularly to Ms. Varsha Prabakaran and Mr. Holger Schäpe, who supported this publication.

Gheorghe I. Gheorghe
Editor in Chief and Conference Chairman

Contents

Effect of Vibration Frequency on Mechanical Behavior of Automotive Sensor

Rochdi El Abdi[1(✉)], Julien Labbé[1], Florence Le Strat[2], and Erwann Carvou[1]

[1] Université de Rennes1- CNRS, Institut de Physique de Rennes- UMR 6251, Campus de Beaulieu, 35042 Rennes Cedex, France
relabdi@univ-rennes1.fr
[2] Entreprise Renault, DEA-TCM, 78084 Guyancourt, France

Abstract. Due to repetitive micro-displacements, the fretting phenomenon was defined as an electrical and mechanical degradation of the electrical contact interface in automotive sensors. Commonly, the electrical degradation was quantified by the increase of contact resistance deduced from the contact voltage. This work aims to address the analysis of relative displacements according to three space directions between sensor components in contact for different vibration frequencies for a Top Dead Center sensor. Particular attention was paid to measurements of displacements near crimping zones.

Keywords: Fretting corrosion · Automotive sensors · Relative displacements Vibration frequency

1 Introduction

In the automotive fields, the vehicle vibrations induce relative movements for hundreds of sensors which were located near the engine, inside the seat and in many other electronical components. The vibrations could induce a displacement between two sensor components in contact i.e. the male part and the female part (the pin and the clip) and could generate an electrical failure due to the well-known fretting-corrosion phenomenon [1, 2]. A relative displacement of 5 μm was enough to produce remains at the interface between the pin and the clip and set an intermittent failure at the interface [3]. This phenomenon represents 60% of electrical failure within cars [4]. Electrical contacts were generally made of a substrate of copper alloy plated with a thin protective layer of non-noble metals. Tin was usually used as a protective layer of the substrate in order to combine a good conductivity, good reliability and a low cost. A pure tin was malleable and reacts with the oxygen to give hard and brittle remains which cause high surface damages. The substrate could be reach and it generates copper oxide remains at the contact surface [5].

In the electrical contact field, the fretting-corrosion was an irreversible degradation which avoids a good current conduction. Several analyses were performed to understand this phenomenon and how the current was conducting through the interface for static and dynamic contacts [6, 7]. Due to the many mechanical parameters (forces,

G. I. Gheorghe (Ed.): ICOMECYME 2018, LNNS 48, pp. 1–7, 2019.
https://doi.org/10.1007/978-3-319-96358-7_1

materials used, type of vibration…), the numerical simulations provide interesting results but cannot predict the real life time of a sensor submitted to a fretting-corrosion phenomenon [8].

The aim of this work was to understand the influence of the vibration frequency on the mechanical wear for different sensor parts in contact and to study the relative displacements in three directions for a sensor submitted to a real vibratory profile used in automotive applications.

2 Sensor Used

In automotive applications, the TDC sensor (Top Dead Center), also known as the "speed sensor", is an electrical component whose purpose is to inform the engine management system about the position of the engine piston at the neutral point, as well as the rotation speed of the crankshaft. It's an inductive and an "active" sensor because it creates its electrical signal independently of any outside power source.

The TDC sensor was submitted to vertical and horizontal vibration (Fig. 1).

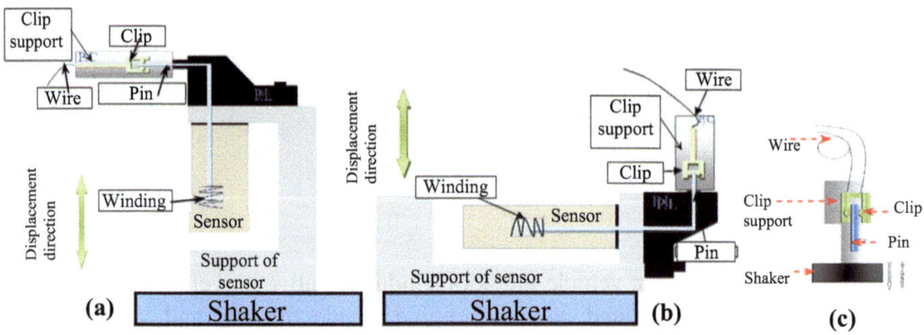

Fig. 1. Schematic representation of TDC sensor in vertical (a) and horizontal (b) position and zoom (c) of clip/pin contact zone

3 Applied Vibratory Profile

The sensor was subjected to vibrations transmitted by the engine and the passenger compartment of the car. When mounted near the car engine, the sensor was submitted to an acceleration and displacement profile given in Fig. 2. This profile will be applied during the tests for studied sensors in our laboratory. The sensor was submitted to displacements between a few micrometers and a few millimeters (Fig. 2).

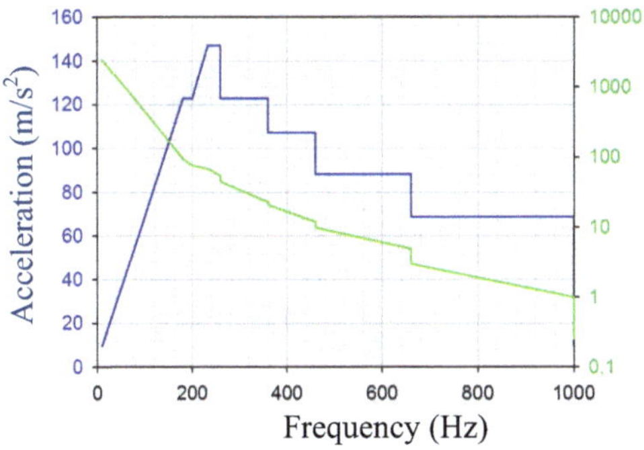

Fig. 2. Vibratory profile used on shaker

On the other hand, the relationship between the frequency *f*, the displacement amplitude *d* (peak-to-peak) and the acceleration γ of a system subjected to vibrations like a shaker system are represented by the following equation:

$$d = \frac{\gamma}{(4.\pi.f)^2} \quad (1)$$

Thus, the connectors were submitted to amplitude of vibrations which depended on the frequency and on the shaker acceleration.

4 Benches Used

The experimental set-up used consists of a shaker (LDS V555 with a maximum force of 939 N, maximum acceleration of 100 g (g = 9.81 m/s^2)) which can apply sinusoidal vibrations and to use the vibratory profile of Fig. 2. A sinusoidal and vertical vibration was applied on the lower part of the sensor which will be positioned horizontally and vertically (Fig. 3).

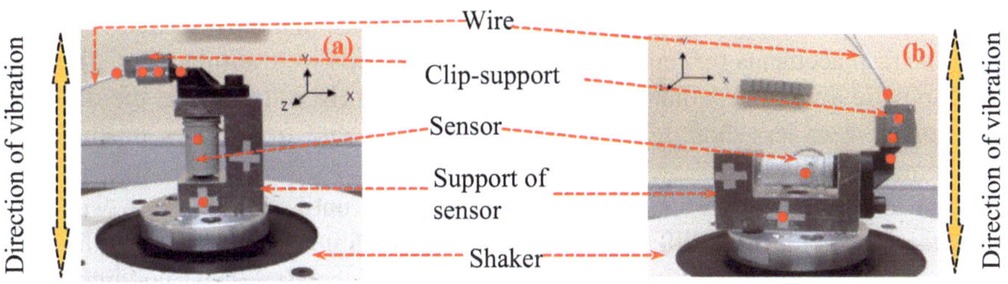

Fig. 3. Shaker used and TDC sensor in vertical (a) and horizontal (b) position. The six red points are targets for laser beam to obtain displacement measurements

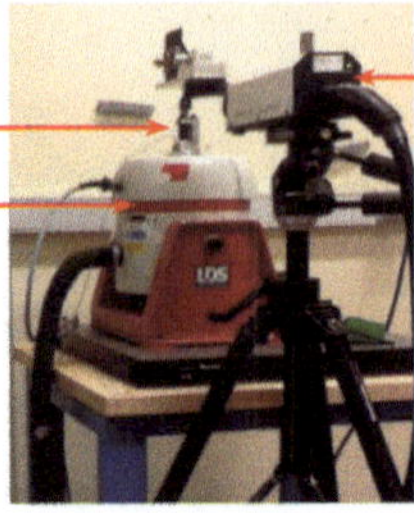

Fig. 4. Laser Doppler Vibrometer used for displacement measurements

A LDV (Laser Doppler Vibrometer) was used to obtain the displacement of target points on the sensors' surface (red points) (Fig. 3). The laser beam from the LDV was directed at the surface of interest, and the vibration amplitude and frequency were extracted from the Doppler shift of the reflected laser beam frequency due to the motion of the surface (Fig. 4). The output of an LDV was generally a continuous analog voltage that was directly proportional to the target velocity component along the direction of the laser beam.

The test aim was to measure for different frequencies the relative displacements between the different sensor parts, especially near crimping zones (contact between wire and sensor components) and at the contact interface between the male part (pin) and the female part (clip) of the sensor (Fig. 1).

5 Discussion and Results

(a) Displacements of Sensor in Vertical Position

Sensor was positioned in vertical position and submitted to vertical shaker vibration (Fig. 3a). From 240 Hz to 440 Hz, the displacements along X-axis of the clip-support, the clip and the wire are similar (Fig. 5a). These displacements decrease according to the frequency and increase for a frequency higher than 600 Hz. The clip-support always has a greater displacement range than the clip and the sensor. The movements of the free wire influence the displacements of the connector as a whole. Note that the clip holder is attached to the sensor by a nut and this causes multidirectional movements.

All the displacements (except for the wire) along Y axis were similar (Fig. 5b), with low amplitude always less than 0.35 µm. Between 200 Hz and 740 Hz, the displacement ranges were less than 1 µm. The wire with one end free causes much higher displacement amplitudes.

When analyzing vibrations along Z axis (Fig. 5c), until 200 Hz there were few differences between the different displacement amplitudes, but these differences become more and more important when the frequency increases. The maximum value of the relative displacement between the wire and the sensor was less than 7 µm.

(b) Displacements of Sensor in Horizontal Position

The sensor was horizontally placed on the shaker (Fig. 3b). From 120 Hz to 800 Hz, the displacements along X axis (Fig. 6a) become high and reach a maximum value when the frequency reaches 680 Hz. These displacements were smaller than those obtained for Z axis where all displacements are similar (Fig. 6c).

Along Y axis (Fig. 6b), the curves are similar, but with a phase shift and the differences between different displacements were always less than 1.3 μm except for the wire with high displacement amplitude at 800 Hz.

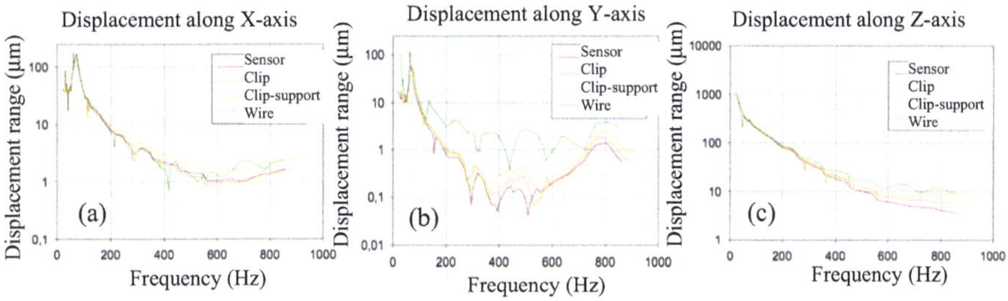

Fig. 5. Displacements of sensor in vertical position along X, Y and Z axes

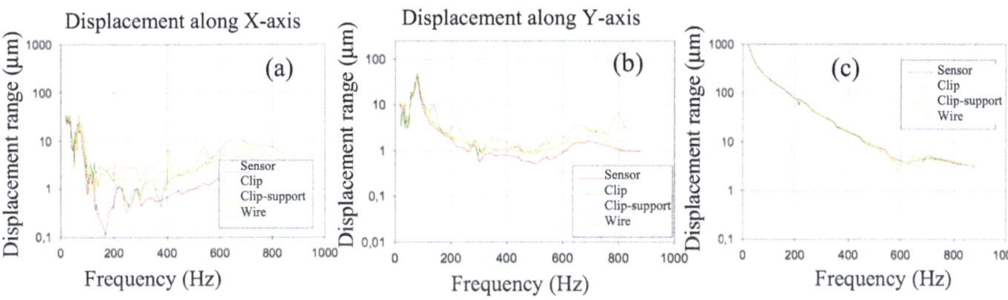

Fig. 6. Displacements of sensor in horizontal position along X, Y and Z axes

(c) Relative Displacements Between Clip-Support and Sensor

An interest was placed on the analysis of the relative displacements between the clip support and the sensor and between the clip and the pin (Fig. 1).

On the other hand, for a sensor horizontal position (Fig. 7b), the relative displacement along Z axis exceeds 20 μm for the frequency of 200 Hz. This leads to a sudden increase of the sensor electrical voltage. This frequency should therefore be avoided. The same conclusions for the relative displacements between the clip and the pine were obtained (Fig. 8).

The damaged area of the contact zone has 300 μm in long (between 200 μm and 500 μm, (Fig. 9)) and the copper appears. At the left of this area, much debris were ejected. In the middle of the contact surface, the tin layer no longer exists and thus the copper will oxidize and lead to the sensor dysfunction.

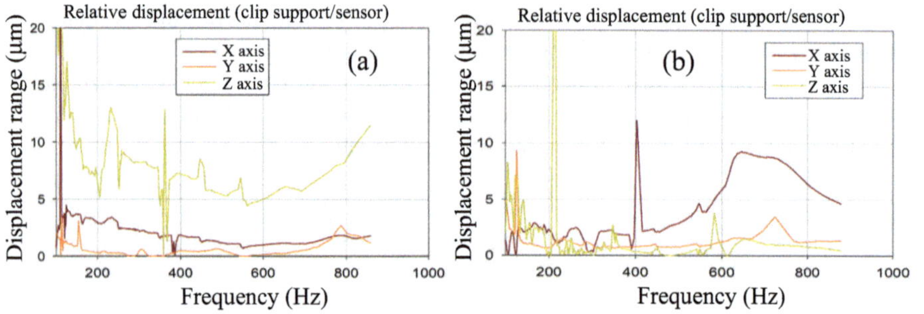

Fig. 7. Relative displacements between clip support and sensor along X, Y and Z axes for sensor in vertical (a) and horizontal (b) position

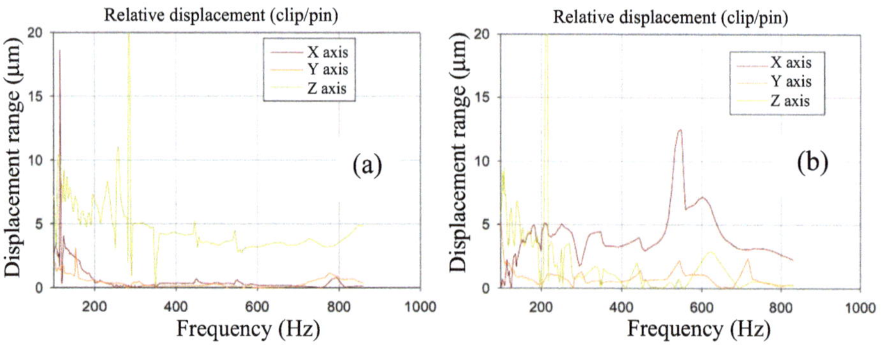

Fig. 8. Relative displacements between clip and pin along X, Y and Z axes for sensor in vertical (a) and horizontal (b) position

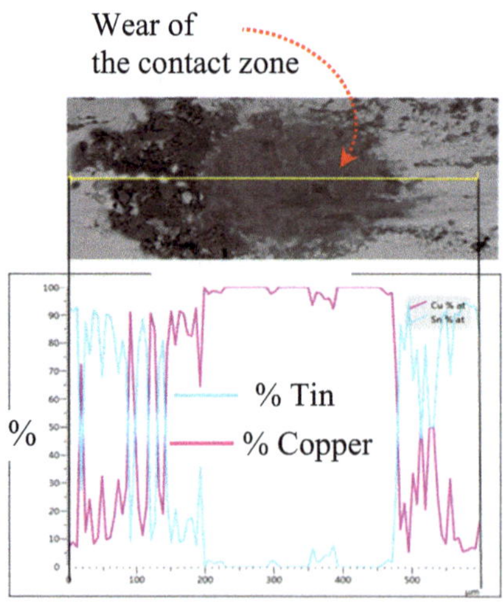

Fig. 9. SEM analysis and material percentages along contact zone between clip and pin

6 Conclusion

The use of a laser Doppler vibrometer with non-contact vibration measurements allowed defining the type of relative movements between different components of sensor used for automotive applications. This has made it possible in particular to emphasize non-intuitive vibratory behaviors such as multiaxial movement directions.

The displacement amplitudes of the clip-support generally were greater than those of the other components of the sensor subject to the shaker vibratory. Indeed, the clip-support was not perfectly fixed to the sensor. Moreover, the results showed the clip-support vibrating were tridimensional even if the shaker vibration was unidirectional. Therefore, it was necessary to characterize the vibrational behavior of each component in three directions.

On the other hand, the vibratory behavior of the clip-support slightly influences the clip movement inside the sensor.

Finally, a sensor subjected to vibration excitations will have a multiaxial vibratory behavior which depends on the imposed vibrations and on the sensor type. Indeed, the vibratory behavior of the each sensor component depends on those of the other external components. However, they may be different. Therefore, analyzing the vibrational behavior of a connector was complex and requires a complete analysis.

References

1. Chen, C., Flowers, G.T., Bozack, M., Suhling, J.: Modeling and analysis of a connector system for prediction of vibration-induced fretting degradation. In: IEEE Holm Conference on Electrical Contacts, pp. 129–135 (2009)
2. Labbé, J., El Abdi, R., Carvou, E., Le Strat, F., Plouzeau, C.: Vibration induced at contact point of tighten-up connector system. In: IEEE Holm Conference on Electrical Contacts, pp. 200–204 (2014)
3. Bouzera, A., Carvou, E., Benjemâa, N., El Abdi, R., Tristani, L., Zindine, E.M.: Minimum fretting amplitude in medium force for connector coated material and pure metals. In: IEEE Holm Conference on Electrical Contacts, pp. 101–107 (2010)
4. Stocker, U., Bonisch, G.: ATZ Automobiltech Z **93**, 7–10 (1991)
5. Park, Y.W., Sankara Narayanan, T.S.N., Lee, K.Y.: Fretting corrosion of tin-plated contacts: evaluation of surface characteristics. Tribol. Int. J. **40**, 548–559 (2007)
6. Benjemâa, N., Carvou, E.: Electrical contact behaviour of power connector during fretting vibration. In: IEEE Holm Conference on Electrical Contacts, pp. 263–266 (2006)
7. Jedrzejczyk, P., Fouvry, S., Chalandon, P.: Quantitative description of the electrical contact endurance under fretting condition: comparison between tin and silver. In: IEEE Holm Conference on Electrical Contacts, pp. 272–277 (2008)
8. Tsukiji, S., Sawada, S., Tamai, T., Hattori, Y., Iida, K.: Direct observations of current density distribution in contact area light emission diode wafer. In: IEEE Holm Conference on Electrical Contacts, pp. 62–68 (2001)

Applications of Additive Technologies in Realization of Customized Dental Prostheses

Edgar Moraru[1(✉)], Daniel Besnea[1], Octavian Dontu[1], Gheorghe I. Gheorghe[2], and Victor Constantin[1]

[1] Politehnica University of Bucharest, Splaiul Independenţei no. 313, 6th District, Bucharest, Romania
eddy_milan91@yahoo.com, d_bes@yahoo.com, octavdontu@yahoo.com, victor.f.constantin@gmail.com
[2] INCDMTM, Pantelimon Road, no. 6-8, 2nd District, Bucharest, Romania
geocefin@yahoo.com

Abstract. The paper presents 3D DLP (Digital Light Processing) digital printing technology, which belongs to the additive technology category using liquid raw material allowing the obtaining of dental prostheses with complex and detailed geometry, fine and precise printed surfaces, fairly resistant dental crowns, with a variety of resins, including biomedical materials (certified for medical use), characterized by high print speed and productivity for complex geometries and high productivity with numerous applications in the medical field. In the paper, a prototype of a dental bridge was made using 3D DLP printing technology, which can serve as a model for future dental restorations.

Keywords: 3D printing · Photopolymerization · Digital Light Processing Dental prostheses · Additive manufacturing

1 Introduction

Three-dimensional printing is a modern technology with applications in several domains. The flexibility of the 3D printing system allows the use of a variety of materials. Most Additive Manufacturing systems using liquid raw materials use a heat source that scans the 2D surface of the liquid, and at impact with it, produces "point-to-point" or "surface-to-surface" solidification. The piece is built on a horizontal platform submerged in the liquid polymer. The solidification is achieved by the photopolymerization produced at the impact of a light ray with the upper surface of the liquid. Photopolymers are light-sensitive polymeric materials that change their physical or chemical properties when exposed to an external stimulus (heat source, ultraviolet radiation, etc.). The main source of light is UV light, which initiates a reaction and changes the properties of photopolymers.

The photopolymer has special properties in the sense that radiation in the ultraviolet or visible field initiates the polymerization at a depth of several tenths of a millimeter below the surface of the liquid is restricted to a shape corresponding to a cross section of the piece. After solidifying a layer the piece is submerged with a new section thickness

G. I. Gheorghe (Ed.): ICOMECYME 2018, LNNS 48, pp. 8–17, 2019.
https://doi.org/10.1007/978-3-319-96358-7_2

and a new cross section of the piece is solidified. The entire cycle is repeated until the piece is solidified in its totality [1].

The commonly used photopolymer, the ester of cinnamic acid (C9H8O2)) is produced by reacting cinnamic acid with alcohol in the presence of light [2]. After the piece has been obtained by this process, a post-processing operation is carried out using a light and/or heat source that must be so chosen that it does not destroy or distort the piece (laser, UV, sunlight, etc.). Light from these sources initiates chemical reactions that change their structure and change their chemical and mechanical properties. The fact that photopolymerisation occurs only a few tenths of a millimeter thick on the surface of the liquid allows the use of less translucent polymers and lower light intensity. Some of the common polymeric bases are polyvinyl cinnamate, polyamide (PA), polyisoprene, polyimides, epoxides, acrylics etc. Usually with these polymers, monomers, oligomers and additives are used [3].

Binders are reactive intermediate molecular weight molecules consisting of several monomer units, usually dimers (two units), trimers (three units) and tetramers (four units). They are normally liquid at room temperature and are used as ink, adhesives and coatings. Typical photopolymers consist of 50–80% of such binders/oligomers, some examples of binders:

- Styrene family: Oligomer Of Styrene-Tetramer-Alpha Cumyl End Group, A-Methyl

  Styrene-Dimer (1), A-Methyl Styrene-Tetramer etc.
- Methacrylate family: Acrylic Acid Oligomers, Methyl Methacrylate Oligomers, Methyl

  Methacrylate Tetramer etc.
- Vinylalcohol family: Vinyl Alcohol Trimer, Vinylacetate Trimer, Vinylacetate Oligomer
- Olefine family: Poly Isobutylene
- Glycerol family: Triglycerol
- Polypropylene Glycol Family: Poly Propylene Glycol (Dihydroxy Terminated) etc.

Monomers are small chemically bound molecules that repeatedly bind to other monomers, oligomers or polymers to form new polymers. Mostly photopolymers consist of monomers based on acrylates or methacrylates ranging from 10 to 40%. In the polymerization process, two types of monomers can be used: multifunctional monomers and monofunctional monomers. Multifunctional monomers can act as both diluents and crosslinkers, while monofunctional monomers may be either diluents or crosslinkers (Table 1).

Table 1. Monofunctional and multifunctional monomers

Monofunctional monomers	Multifunctional monomers
Acrylic acid	Trimethyllopropane triacrilate (TMPTA)
Methacrylic acid	Ethoxylated TMPTA
Isodecyl acrylat	Trymethyllepropane trimethacrylate
N-vinyl pyrrolidone	Hexanediol diacrylate

The photoinitiators convert light energy into chemical energy by forming free radicals or cations upon UV exposure. They can break into two or more particles with UV reaction and at least one of the particles will react with monomers or oligomers and binds them together. They exist naturally or can be chemically synthesized and are sensitive to specific wavelengths of light. The photoinitiators form only a small component of photopolymers [2] (Table 2).

Table 2. Examples of photoinitiators

Free radical photoinitiators	Cationic photoinitiators
Isopropylthioxanthone	Diaryliodonium salts
Benzophenone	Triarylsulfonium salts

In the photopolimerization of free radicals, the radicals or ions break off the initiators when UV reacts and the ions will begin to react with monomers to initiate the polymerization. During the cationic reaction, the strong acid will be released from the initiator and a binding process begins. Diaryliodonium and Triarylsulfonium salts are stable crystalline compounds which can be prepared by synthetic method and are available in the market for commercial purpose [2] (Fig. 1).

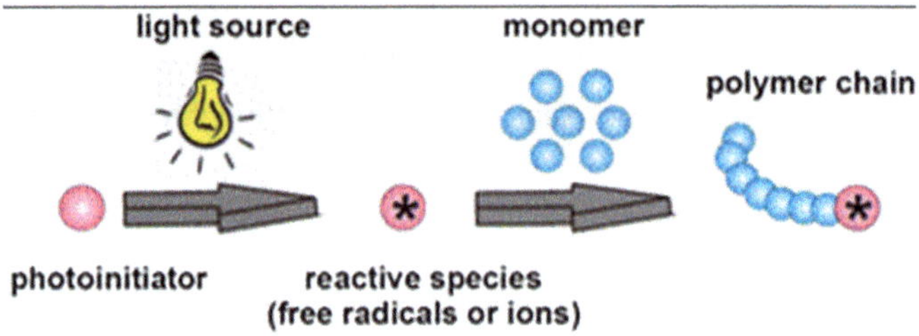

Fig. 1. Photopolymerization [12]

Our teeth play a very important role in our daily lives. They help not only to crush the food, but also to improve the digestive process. Moreover, they are aesthetically important. We can conclude that the teeth should be in a complete set. Only in these circumstances a person will be able to maintain his health and attractiveness [4]. Modern dentistry has taken the best ideas since antiquity and has multiplied it into today's technologies and materials. As a result of this symbology, the dental prostheses of today's days have appeared, which are worthy of admiration. They are dental structures designed to restore the anatomy and physiology of the dentition. Most often, the prostheses should be installed if one or more teeth have been lost. A person may lose teeth because of various diseases (such as dental caries and periodontitis). In addition, the cause of the loss may be the lesion or malformation of the dental system. The installation of dental prostheses allows restoration of mastication, speech, avoidance of teeth movement. Almost all modern prostheses are very close to the properties and aspect of the natural teeth. The actual manufactured prostheses are comfortable, practical and with proper care they can serve for a long time [5, 6] (Fig. 2).

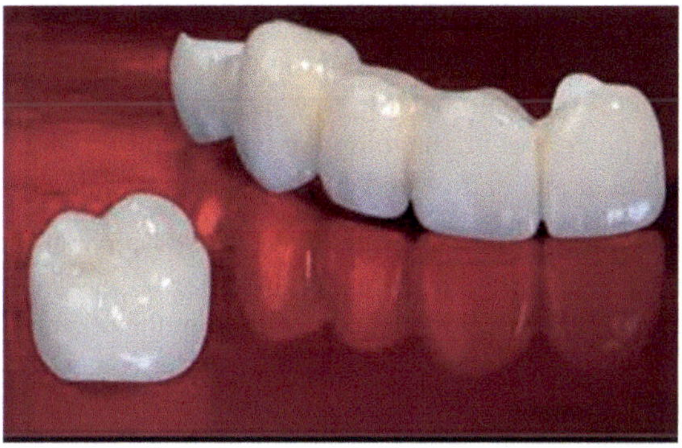

Fig. 2. Ceramic dental crown [7]

Dental prostheses can be classified from several points of view: by mobility (removable, mobile, and fixed), depending on the aesthetic aspect (physiognomic, semi-physiognomic, non-physiognomic), depending on the material used (metallic, ceramic, metal-polymeric, metal-ceramic, etc.).

This branch of prosthetic dentistry is constantly developing. More and more often we hear from different sources about the possibilities of a new 3D printing technology [13]. After all, if it is possible to print any three-dimensional figure, why would not it be possible to print a tooth? Compared to milling, everything is exactly the opposite, we do not eliminate too much, but the layer with layer is formed or "imprinted" the future prosthesis according to the 3 D CAD/CAE model. The fabrication of prostheses using CAD technologies allows providing support for each irregularity of the later metal-ceramic restoration, since it first models the anatomical shape and then using the programs, uniformly removing the ceramic application layers. This ensures the stability of the restoration and a long lifetime of the construction [14] (Fig. 3).

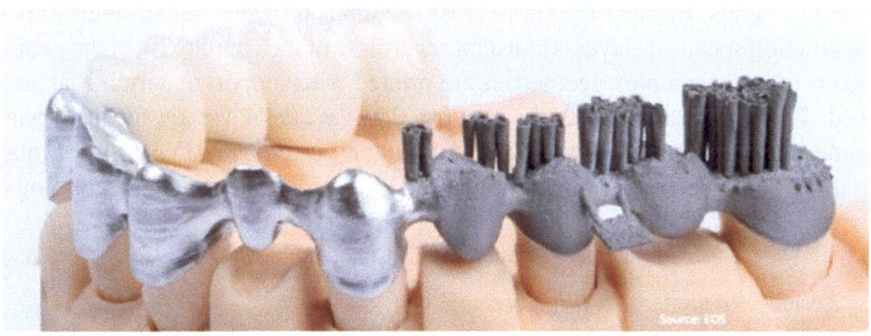

Fig. 3. Dental prosthesis realized by additive technologies(SLS-Selective Laser Sintering) [15]

An application of great impact and perspective has AM (Additive Manufacturing) technologies in the field of personalized medical implantology. The development of these applications has gone from two premises: the first is that every human is a prototype, and the second, the ability of computed tomography (CT) or nuclear magnetic resonance (PMN) systems to deliver data and information to go end to a virtual 3D model of the area of interest that can be saved as a .stl file, specific to the fabrication technology by adding material. From this moment on, the physical materialization of the virtual model of interest in a process of implantology or bone reconstruction is treated like any other piece. A physical model in this field can have several utilities:

- He can serve in the proper diagnosis and selection of suitable therapy, improving communication between different groups of doctors;
- The three-dimensional physical model can be used in the planning of very complex surgical interventions that can be "practiced" on these models. For example, in maxillofacial surgery for osteotomy planning, for correct choice of bone sectioning paths and how to rearticulate them;
- Using a physical model makes it easier and more precise to realize a prosthesis or implant;
- It can even be a personalized implant made of biocompatible material for the patient;

Applications of AM technologies in this area are gaining more and more importance. Challenges are related to the finding and approval of new biocompatible materials that can be put into the desired physical form, with physico-mechanical structures and properties as close as possible to those of the bone part it replaces [1].

2 Materials and Methods

DLP Printing Technology (Digital Light Processing) is an additive manufacturing process based on the use of UV light for the solidification of liquid polymer resins. DLP technology has as its main element the DMD (Digital Micromirror Device) chip - a matrix of micro-mirrors used for fast spatial light modulation. Each individual micro-mirror of the DMD chip projects pixels from the cross-section of the 3D model. Under the action of UV light, the photoreactive liquid (sensitive to ultraviolet light) solidifies in successive layers. Because the entire cross section is projected into a single exposure, the construction speed of a layer is constant regardless of the complexity of the geometry. 3D objects of more complex geometries are printed with support materials that are later removed. The resin remaining in the construction vat can be reused for later printing. For some printed materials, subsequent curing processes in UV furnaces may be required. Table 3 lists some of the features of Digital Light Processing technology.

Table 3. Features of DLP technology

Accuracy of printed parts	Very good
Finishing printed surfaces	Very good
Print speed	Good (for multiple objects and complex geometries)
Used materials	Resins, photopolymers, transparent resins

Advantages of DLP (Digital Light Processing) technology: Fine and precise printed surfaces (use in the jewelry industry, dental technology, electronics), the prototypes is quite resistant for processing, a variety of resins including bio-medical materials (certified for medical use) and transparent resins (prototypes in the packaging industry), stable printers with few moving parts. The technology allows prototyping of parts with complex and detailed geometries, high printing speed for complex constructions and simultaneous printing of many pieces (high productivity). Printed parts can be used as master molds for injection molding, thermoforming, metal casting.

Disadvantages of DLP (Digital Light Processing) technology: Expensive building materials, higher 3D printer prices (for large volumes), require post-processing operations (UV curing, removal of support material) require handling of resins.

DLP (Digital Light Processing) technology applications: Resistant prototypes for functional testing, fine prototypes and models, precise prototypes with complex geometries; fabrication of small series of models in medicine (dental models, dental implants, cochlear implants, dental restorations, medical implants), media models (animation, cinema), jewellery casting models, tools and instruments, parts and components in automotive and aerospace industry [8–11] (Fig. 4).

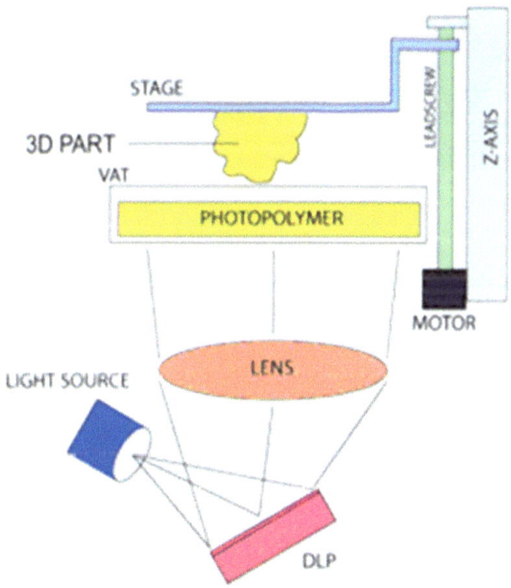

Fig. 4. DLP Digital Light Processing printing technology

3 Experimental

The experimental research consists of creating a dental prosthesis with a 3D Duplicator 7 printer that uses the direct impact of photosensitive resin with UV light transmitted through an LCD screen, Fig. 5.

Fig. 5. 3D duplicator printer 7, general view; 1 - 3D duplicator; 2 - PC

It starts from a CAD model that is saved in .stl format and can be processed by the 3D Duplicator 7 software (Fig. 6).

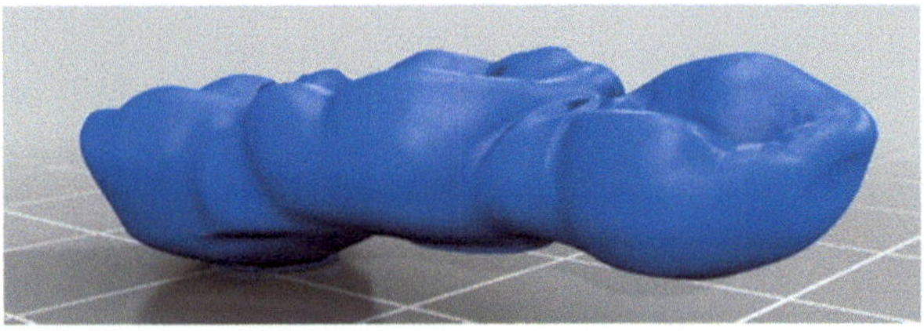

Fig. 6. Proteza dentara CAD

The Creation Workshop software allows the visualization and position of the workpiece on the work plane as well as the mirroring operations. The interface is shown in Fig. 7.

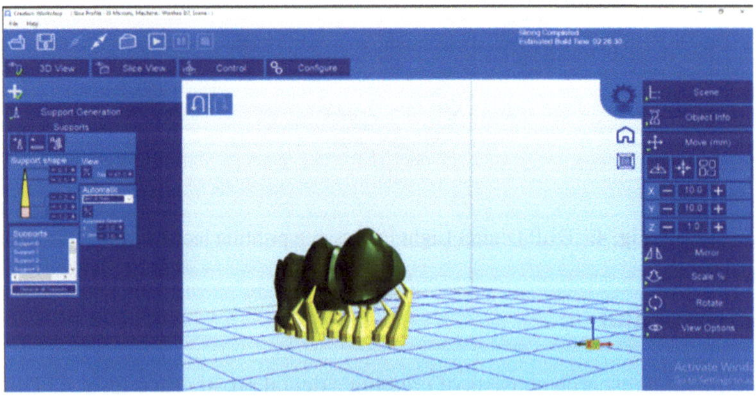

Fig. 7. Program interface

Dental prosthesis from the CAD model does not have a laying surface, in this case it is necessary to add a sacrificial layer, as shown in the figure above.

The technical specifications of the 3D Duplicator 7 printer are shown in Table 4:

Table 4. The technical specifications of the 3D duplicator 7 printer

Technical specifications of the 3D duplicator printer 7		Resin specifications	
Print volume	2,1 l	Wavelength	405 nm
Print sizes	120 × 68 × 200 mm	Solidification time	8–16 s
Thickness layer	min. 0,035 mm	Viscosity (at 25 °C)	120–140 mPa * s
Print speed	30 mm/h	Density (at 25 °C)	1,12 g/cm^3
Resolution XY	2560 × 1440 px	Hardness (XY)	79,2 MPa
Software	Creation workshop	Hardness (Z)	73,5 MPa
		Flexural strength (XY)	8,43 MPa
		Flexural strength (Z)	5,57 MPa
		Tensile strength (XY)	21,4 MPa
		Tensile strength (Z)	15,2 MPa

While FDM technology produces a mechanical bond between layers, DLP (Digital Light Processing) technology creates a chemical bond by bonding the photopolymers through layers, resulting in very dense parts, the bond is water and air tight, and the resistance does not changes according to orientation (Figs. 8 and 9). The experimental results led to the realization of a prototype of the dental prosthesis which is shown in Fig. 10.

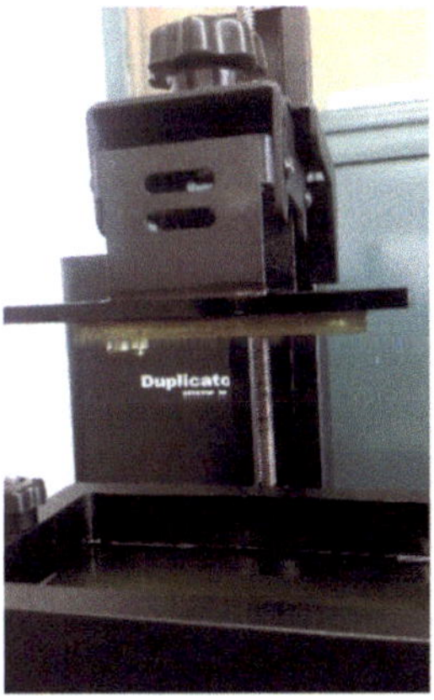

Fig. 8. 3D duplicator printer 7 [8]

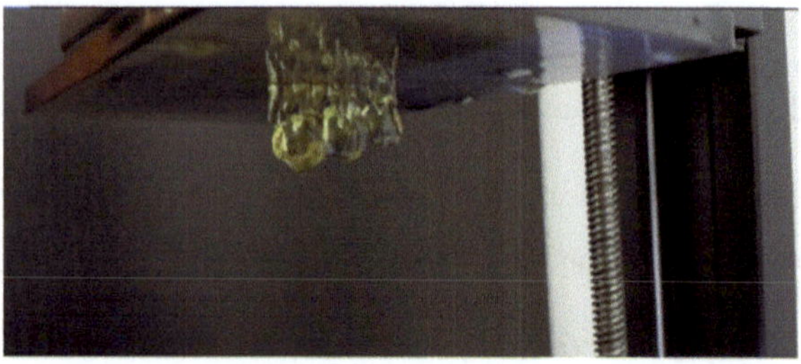

Fig. 9. Dental prosthesis on the work platform

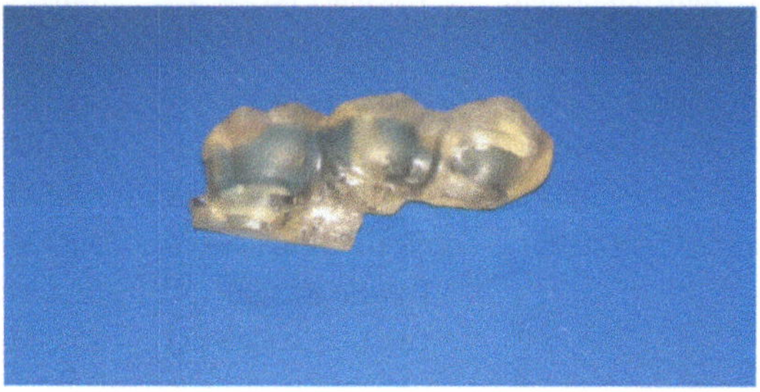

Fig. 10. Dental prosthesis obtained with DLP digital light processing technology

In about two and a half hours a prototype of dental prosthesis has been obtained that can be a landmark for the dentist in order to obtain the future prosthetic restoration.

The sacrificial layer was later removed, the prosthesis being cleaned in a 92% technical alcohol bath. After this, the restoration was left for 24 h in a water recipient to complete the photopolymerization.

4 Conclusions

The development of photopolymers with desired properties such as chemical composition, mechanical, high degree of biocompatibility will be useful in the creation of complex products and thus increase the application areas, especially in the medical field.

Tooth loss due to various pathologies at present is considered to be a serious health problem. In this case, not only the aesthetic and phonetic roles intervene. Perhaps the most important function of the teeth is mastication, thus improving the digestive process by shredding the food. Because of this the installation of dental prostheses that replace the lost or damaged teeth is absolutely indispensable.

Throughout the existence of dentistry science, it has appeared a wide range of procedures and techniques for replacing dental defects. They are constantly improved along with the evolution of the scientific base, and at the same time new methods, materials and technologies are found to minimize or eliminate the inherent shortcomings of a treatment method. Lately, the additive technologies have gained much ground regarding the realization of dental prostheses, gradually replacing conventional casting technologies. These technologies provide an efficient and rapid method for designing and manufacturing biocompatible metal, ceramic and polymeric carcasses for complex dental frameworks. The selective laser sintering is one of the additive technologies that result in dental crowns with high performance features. The unused material can be used in the following processes, making this technology more efficient from economical point of view than CAD milling, the other completely opposite technology for creating prostheses.

In the coming years, notable achievements will be highlighted in this field, both in terms of new biocompatible materials and the ability of AM - Additive Manufacturing technologies to achieve dental prostheses with variable structures.

This study highlighted the realization of prototypes of dental prostheses using Digital Light Processing technology, yielding admirable results that can be used as models for future customized dental crowns or even as temporary prostheses.

Acknowledgements. This work has been funded by University POLITEHNICA of Bucharest, through the "Excellence Research Grants", Program UPB-GEX 2017. Identify: UPB-GEX2017, Grant no. 48/25.09.2017, ME 14-17-05, ID98.

References

1. Berce, P., Balc, N., Caizar, C., Pacurar, R., Radu, A.S., Bratean, S., Fodorean, I.: Tehnologii de fabricatie prin adaugare de material si aplicatiile lor, Editura Academiei Romane, Bucuresti (2014)
2. Pandey, R.: Photopolymers in 3D printing applications, Degree Thesis Plastics Technology (2014)
3. Sheridan, J.T.: Photopolymers materials (light sensitive organic materials): characterization and application to 3D optical fabrication and data storage (2014)
4. http://worlddent.ru/zubnye-protezy
5. http://www.rusmedserv.com/toothreplacement/denture/
6. http://www.lacalut.ru/information/articles/view/185-fixation-prostheses
7. https://www.mycostamesadentist.com/blog/choosing-your-dental-crown/
8. www.wanhao3Dprinter.com
9. 3D Printing with Desktop Stereolithography, An Introduction for Professional Users, June 2015. formlabs.com
10. Xu, Y., Wu, X., Guo, X., Kong, B., Zhang, M., Qian, X., Mi, S., Sun, W.: The boom in 3D-printed sensor technology. Sensors **17**, 1166 (2017). https://doi.org/10.3390/s17051166
11. Gross, B.C., Erkal, J.L., Lockwood, S.Y., Chen, C., Spence, D.M.: Evaluation of 3D printing and its potential impact on biotechnology and the chemical sciences. Am. Chem. Soc. (2014). https://doi.org/10.1021/ac403397r
12. http://web.itu.edu.tr/~yusuf/index-1.htm
13. http://www.print3dbucuresti.ro/aplicatii-printare-3D/printare-3d-tehnica-dentara/
14. Майстренко, А.А., Толчек, Л.Г.: – Компьютерные технологии в стоматологии
15. https://www.sinterex.com/metal-3d-printing-and-additive-manufacturingblog/2016/6/13/see-how-dentistry-is-benefiting-from-metal-3D-printing

Researches Regarding the Use of Additive Technologies in the Construction of Water Aeration Elements

Octavian Dontu[1(✉)], Moga Corina Ioana[2], Beatrice Tănase[1], Nicolae Băran[1], Gheorghe I. Gheorghe[3], and Edgar Moraru[1]

[1] Politehnica University of Bucharest, Splaiul Independenţei no. 313, Sector 6, Bucharest, Romania
octavdontu@yahoo.com, tanaseelenab@gmail.com, n_baran_fimm@yahoo.com, eddy_milan91@yahoo.com

[2] D.F.R. Systems SRL Bucharest, Bucharest, Romania
corinamoga@yahoo.com

[3] The National Institute of Research and Development in Mechatronics and Measurement Technique, Pantelimon no. 6-8, Sector 2, Bucharest, Romania
geocefin@yahoo.com

Abstract. The paper presents the technology of execution of a fine bubble generator; the air inlets in the water are Ø 0.1 mm and made in silicon plates. The scheme of the experimental installation where this fine bubble generator is fitted is presented. The end of the paper shows the methodology of the researches and the obtained results.

Keywords: Water aeration · Fine bubble generators · Additive technologies

1 Introduction

In water treatment and purification processes, aeration is the basic operation in ensuring proper water quality. Aeration is used in the following areas:

- in water treatment processes for the removal of dissolved inorganic substances or chemical elements such as iron, manganese, etc., by oxidation and formation of sedimentable compounds or which may be restrained by boiling;
- in the biological treatment of wastewater either by the activated sludge process or bio filters;
- in the processes of disinfection by ozonisation of raw water taken from a source for the drinking purpose;
- in separating and collecting emulsified fats from wastewater.

Water aeration is a mass transfer process with wide application in water treatment. The aeration equipment's are based on the dispersion of one phase into the other, for example, gas into liquid, consuming energy. Dissolved oxygen is an important parameter in assessing water quality due to its influence on living organisms in a volume of water.

G. I. Gheorghe (Ed.): ICOMECYME 2018, LNNS 48, pp. 18–28, 2019.
https://doi.org/10.1007/978-3-319-96358-7_3

A height or low dissolved oxygen level can affect water life and quality [1]. Non-compound oxygen, or free oxygen (O_2), is the oxygen that is not bound to any other element (Fig. 1). Dissolved oxygen is the presence of those free O_2 molecules in water. The water-bound oxygen molecule (H_2O) is a compound and is not considered in the determination of the dissolved oxygen level [2].

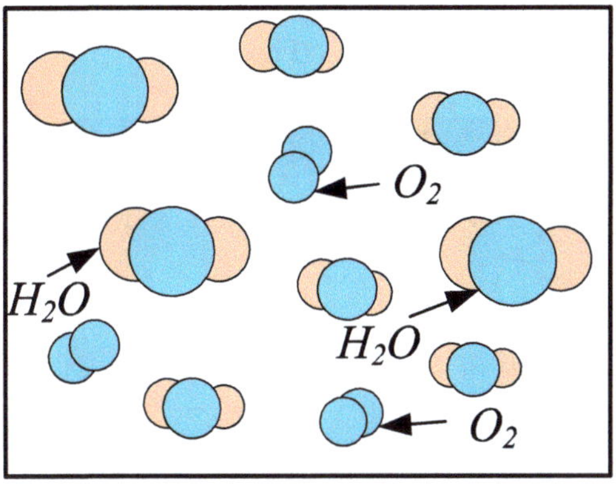

Fig. 1. Dissolved oxygen in water

In Fig. 1 one can observe that oxygen appears in two forms:

- O_2 bounded to H_2;
- free O_2 called dissolved oxygen in water;

The solubility of oxygen in water depends on the temperature, pressure, the size of the air-water contact surface and its turbulence. Any type of water source has its own physic-chemical and biological characteristics and varies from one region to anther depending on the mineral salts composition of the covered areas, contact time, temperature, climate conditions.

Dissolved oxygen is required for many life forms, including fish, invertebrates, bacteria and plants. These organisms uses oxygen in breathing. Fishes and crustaceans obtain oxygen for breathing through their gills, while plants and phytoplankton need dissolved oxygen for breathing, when there is no light for photosynthesis [3]. The amount of dissolved oxygen required varies from life to life. Crabs, oysters and worms require minimal amounts of oxygen ($1 \div 6$ mg/dm^3), while fishes in shallow water need a higher level ($4 \div 15$ mg/dm^3) [4, 5]. Dissolved oxygen enters into water from the air or as a by-product of the plant. Oxygen can diffuse slowly on the surface of the surrounding water, or it can be rapidly mixed by aeration, either natural or man-induced [6]. Water aeration can be caused by wind (creating waves), waterfalls or other forms of running water. There are a variety of water aeration modes [1–3].

They fall into the following categories:

a) natural aeration;
b) surface aeration;
c) underground aeration.

There are a number of techniques and technologies available for the three categories [2, 3].

2 Presentation of the Fine Bubble Generator

In a fine bubble generator (FBG), the air dispersing element in water may be a flat plate with orifices (Fig. 2) or a disc with orifices (Fig. 3).

Fig. 2. Processed plate

Fig. 3. Disc with orifices

The essential element at a FBG is the disk with orifices (Fig. 3) which has to fulfill the following conditions [7]:

(a) to be resistant to the action of the liquid it comes into contact with (waste water, water of a particular pH);
(b) to allow a uniform air distribution with low pressure losses;
(c) to be easily processed;

(d) to exhibit a high mechanical strength sufficient to withstand the weight of a water column of 4 ÷ 5 m.

The assessment of the air pressure loss through the perforated disk is as follows:

The orifices in the perforated disk will be assimilated with n parallel capillary tubes. The pressure loss for an orifice will be calculated as follows [8, 9]:

$$\Delta p = \sum \xi_i \rho_a \frac{w^2}{2} \quad (1)$$

Relation where the sum of the local pressure loss coefficients is made up of sudden section variations [10]:

- the entrance into the orifice: $(s > 3d_0){:}\xi_i = 0.5$
- the exit from the orifice:$(s > 3d_0){:}\xi_i = 1.0$.

So $\sum \xi_i = 1.5$.

The air density will be calculated considering an overpressure given by the height of the water in the tank (h). For an orifice with diameter d_0 an airflow through the orifice results:

$$w = \frac{\dot{V}}{A} = \frac{\dot{V}}{\frac{\pi d_0^2}{4}} = \frac{\dot{V}}{\pi d_0^2}[m/s] \quad (2)$$

where $\dot{V}$ is the air flow measured at a rotameter and divided by the number of orifices.

Figure 4 shows the disc with Ø 0.1 mm orifices.

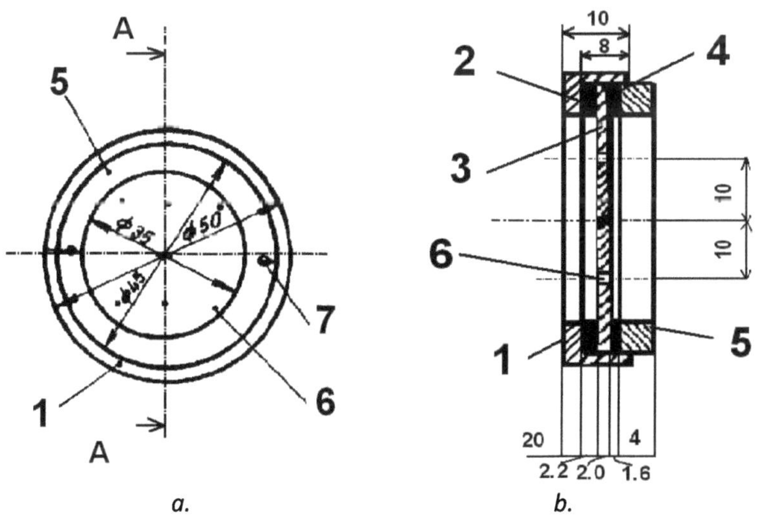

Fig. 4. Box containing the orifices plate: a - view; b - cross Sect. 1 – the box body; 2–gasket; 3–disc with five orifices Ø 0.1 mm; 4–safety ring; 5–nut; 6–orifices; 7–Ø 0.1 mm orifices

3 The Technology for the Execution of the Orifice Disc

The circular plate has orifices obtained by micro processing of mono-crystalline silicon that allows the production of cavities of different shapes, opened or closed orifices of flat/profiled membranes, of elastic elements such as lamellar springs or double plane spirals springs.

The production of these forms is based on wet, selective, anisotropic chemical erosion and with influenced form by controlled doping of the material from which they are made. Selectivity and anisotropy are exploited in the process of manufacturing microstructures in the sense that materials of structuring layers and attack solutions can be "selected" so that the actions are convergent or divergent; based on the recognized anisotropy of the crystalline materials, the corrosion tendency may be accentuated or diminished by the proper orientation of the crystal (Fig. 5).

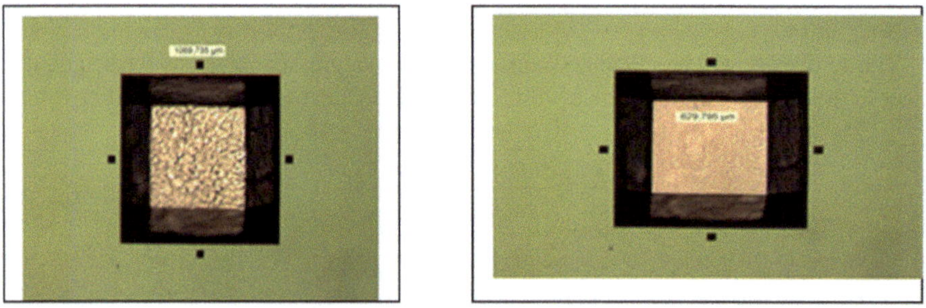

A B

Fig. 5. The structure obtained by chemical erosion on a silicon wafer of 320 μm thickness, face A and B

A 3D printer based on FFF (Fused Filament Fabrication) technology, Fig. 6, was used to construct the support disk with orifices for the silicon plate with the following technical features:

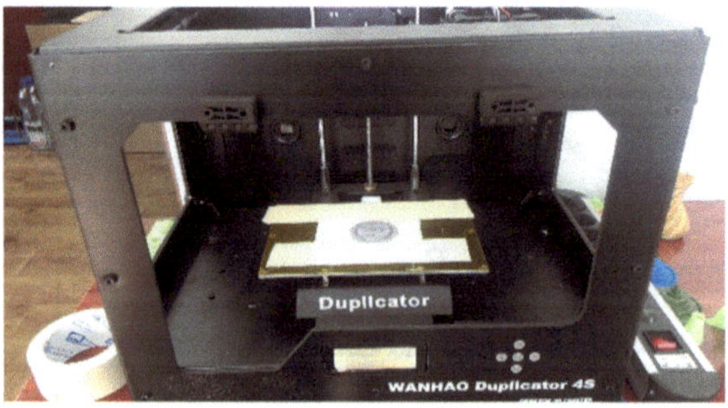

Fig. 6. 3D printer based on FFF (Fused Filament Fabrication) technology

- Print volume: 4.9 l
- Layer thickness: 0.1–0.5 mm
- Z-axis positioning precision = 2.5 μm, X, Y-axis = 11 μm

Processes based on material extrusion use a wire of various materials (PLA, ABS), which heats up to a temperature a few degrees below the melting temperature, then reduces its diameter to 0.12–0.15 mm by extruding it in a depositing device, a device that moves in the XOY plane to materialize a section of the virtual model. This manufacturing process is based on heating the material to be deposited to its melting point and then depositing the melted material where it is needed to build the desired pattern.

The key of the process is to rigorously control the temperature at which the material is heated and maintained during deposition. The heating of the ABS wire is carried out at a temperature of 250 °C or PLA at 215 °C, where the material is in a semi-liquid state and can be extruded through a very small diameter nozzle (0.254 mm or 0.127 mm). The nozzle through which the plastic material is extruded, which is in the semi-liquid state, can be moved together with the heating head on which it is fastened.

This movement is done in the XOY plane, the movement being numerically controlled by a computer. The piece under construction is on a vertically moving platform, along the Z axis, also controlled numerically by the machine control equipment.

In this way, a piece can be made by depositing material where the configuration of the piece in question requires it. It is important that the time of materialization of a virtual model in a physical model is very short, compared to conventional manufacturing processes [7].

In order to increase the productivity, the supports are built with a lamellar structure. In this way, less material is consumed for supports, and they can be more easily separated from the piece material.

The extrusion process (FFF) is widely used in a wide range of industries: automotive, aerospace, consumer products, marketing, medicine, architecture, and so on.

The advantages of the process are given by the possibility of direct realization of the functional parts of ABS and PLA, differently colored materials. Pieces manufactured by FFF systems are also used indirectly as master models for the production of flexible tools for the manufacture of metallic or non-metallic parts in individual or small series production. This is the use of wax models for casting with light fusible patterns, ABS models for the manufacture of rubber molds and metal spraying matrices.

The advantages of FFF systems results from the fact that the manufacturing process does not produce much waste material, it uses cost-effective materials, and FDM systems are easy to use and do not require special installation and operating conditions.

The disadvantages are due to the lower quality of the processed surfaces, mainly because of the scale effect, the lower accuracy (0.1–0.2 mm) and the relatively small size of the parts that can be manufactured.

The steps of the technological process are shown in Figs. 7 and 8 through 3D design using dedicated software package of the lower and upper supports respectively for silicon wafer and saving the .stl files. [11] (Fig. 9).

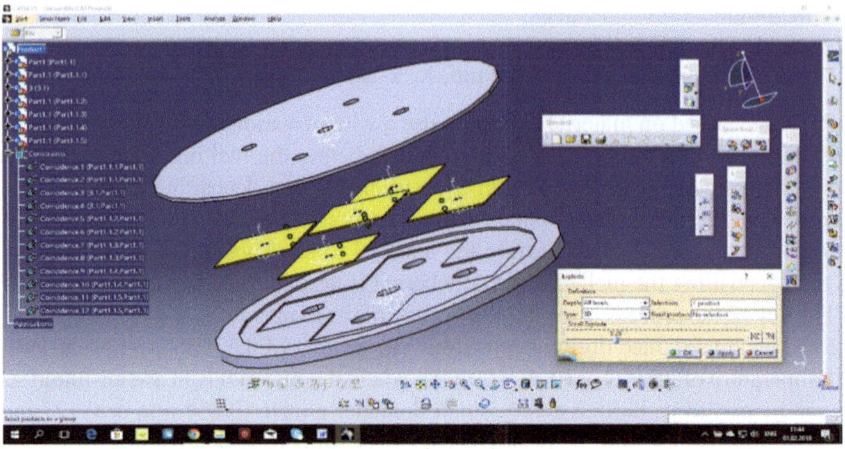

Fig. 7. Designing the CAD model of the fine bubble generator

a

b

c

d

Fig. 8. Support execution for the silicon plates obtained by bulk micromachining technology: a-bottom support plate of the plates obtained by FFF technology; b-bonding the silicon plates to the support plate; c-bonding the top cover; d-capsulation of the fine bubble generator

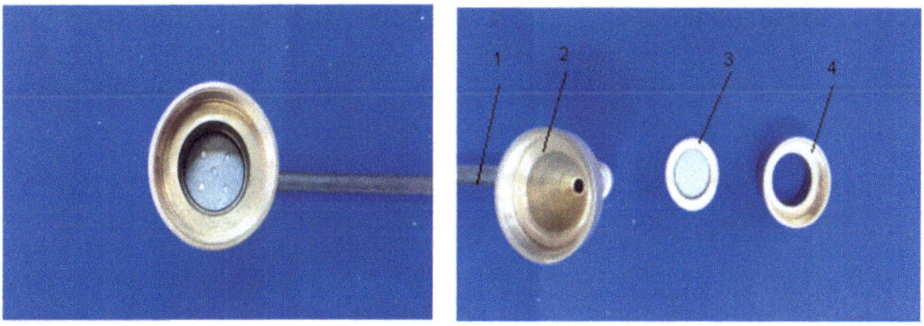

Fig. 9. General view of the fine bubble generator: 1–compressed air supply pipe; 2–the body of the fine bubble generator; 3–the encapsulated fine bubble generator; 4–sealing nut

After the perforated plate is manufactured, it is inserted into the box; the box is mounted inside the FBG and it is fixed to the lower end in the compressed air supply pipe Fig. 10.

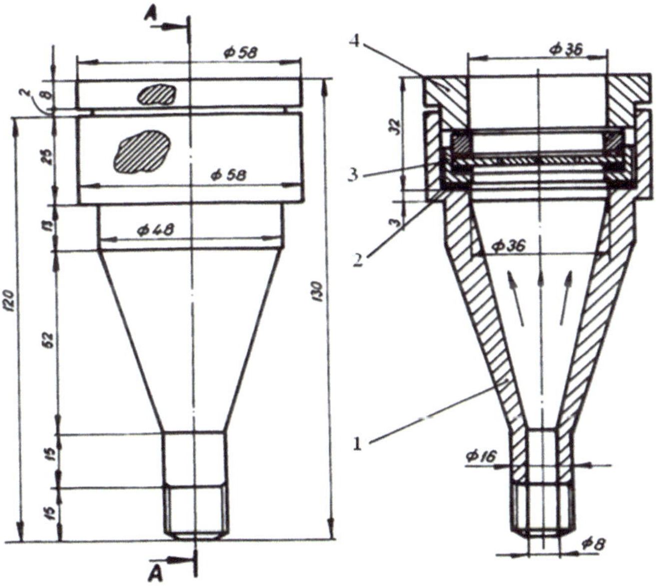

Fig. 10. The fine bubble generator body. a - view; b - cross section; 1–truncated body with threaded bottom; 2–gasket; 3–box; 4–tightening nut

In Fig. 10 for the box (3), the component parts are not detailed because it is represented in Fig. 4. The nut (4) is threaded on the truncated body (1) by tightening it secures the box (3) to the rubber gasket (2) preventing the air from penetrating into the water through orifices other than those in the disk.

4 Experimental Researches on the Fine Bubble Generator

Figure 11 shows the scheme of the experimental installation for the injection of atmospheric air into water.

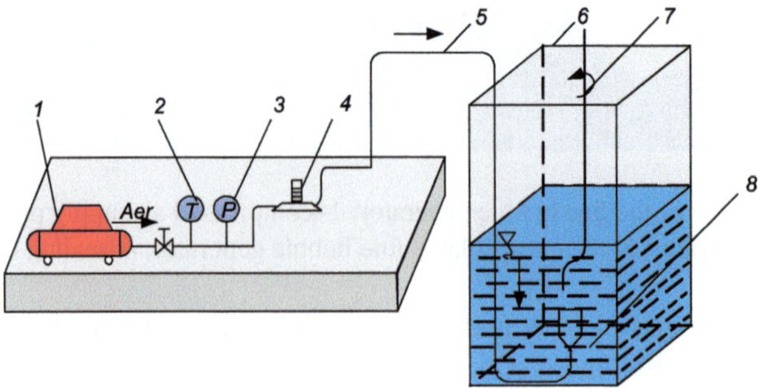

Fig. 11. Sketch of the experimental installation for the injection of atmospheric air into water. 1–air compressor; 2–thermometer; 3–manometer; 4–rotameter; 5–microbubbles generator feed pipe; 6–parallelepiped water tank; 7–oxygenometer probe; 8–microbubbles generator with 152 orifices Ø 0.1 mm

The measurement of dissolved oxygen in water is carried out with an oxygen meter whose operation is based on an electrical method; the oxygen meter is equipped with a probe (7) which during the measurement process performs a rotation movement at a velocity of 0.3 m/s [12].

5 Research Methodology and the Experimental Obtained Results

The measurements involve the following steps:

1. Verification if the 5 orifices are working, i.e. the air is introduced into the fine bubble generator;
2. The tank filling with water up to H = 500 mm H_2O;
3. The measurements of C0, tH_2O, tair;
4. The introduction of the fine bubble generator and the time notation (τ);
5. The removal of the FBG every 15 min outside the tank, and the measurement of the dissolved oxygen concentration.
6. When a horizontal plane of function C = f (τ) is reached, the measurements stops with the condition: C ≈ Cs;
7. From previous research [13, 14], the concentration of dissolved oxygen in water tends to saturation after a two-hour period. So, the oxygen concentration will be measured at times: 15 min; 30 min; 45 min; 60 min; 75 min; 90 min; 105 min; 120 min.
8. At the end of the measurements, the oxygen probe is cleaned and the water from the tank is emptied.

Figure 12 shows a general view of the experimental plant.

Fig. 12. Experimental installation

Figure 13 shows the operation of the fine air bubble generator.

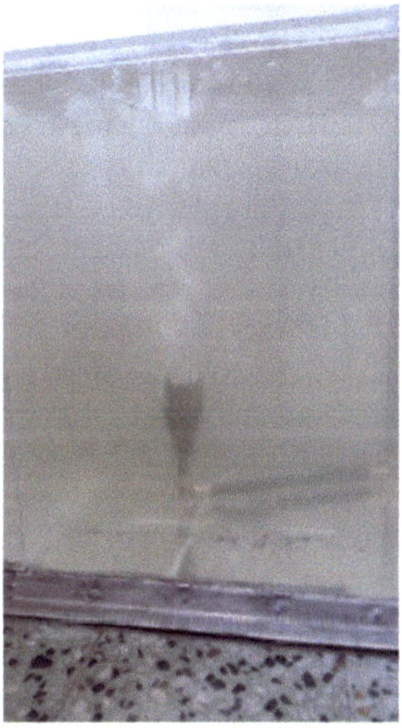

Fig. 13. Fine air bubble generator in operation

From Fig. 13 we one can observe that an air bubbles curtain rises from the FBG body. By using this FBG, within two hours, the dissolved oxygen concentration in water evolves as follows:

- Initial concentration: $C0 = 5.84\ mg/dm^3$;
- Final concentration: $C = 8.39\ mg/dm^3$.

For twater = 23.7 °C the saturation concentration value is $Cs = 8.4\ mg/dm^3$; as a result the experimental obtained results are satisfactory.

6 Conclusions

It is known from the literature [13, 14] that water aeration is even more effective if the diameter of the air bubbles is smaller.

The use of additive technologies solves this problem in the sense that it creates orifices with Ø < 0.1 mm.

Acknowledgements. This work was supported by a grant of the Romanian National Authority for Scientific Research and Innovation, CCCDI – UEFISCDI, project number Manunet - MNET17/ENER2307-CEBIOTREAT, within PNCDI III.

References

1. Pătulea AI., Căluşaru I., Băran N.: Researches regarding the measurements of the dissolved concentration in water. Adv. Mater. Res. **550–553**, 3388–3394 (2012)
2. Oprina, G., Pincovschi, I., Băran, G.: Hidro-gazo-dinamica sistemelor de aerare echipate cu generatoare de bule. Editura POLITEHNICA PRESS, Bucuresti (2009)
3. Oprina, G.: Contributii la hidrodinamica difuzoarelor poroase, teza de doctorat, Facultatea Energetica, Universitatea POLITEHNICA din Bucuresti (2007)
4. Pătulea, Al.: Teza de doctorat, Influenta parametrilor functionali si arhitecturii generatoarelor de bule fine asupra eficientei instalatiilor de aerare, Facultatea de Inginerie Mecanica si Mecatronica, Universitatea POLITEHNICA din Bucuresti (2012)
5. Miyahara, T., Matsuha, Y., Takahashi, T.: The size of bubbles generated from perforated plates. Int. Chem. Eng. **23**, 517 (1983)
6. Căluşaru, I., Băran, N., Pătulea, Al.: The influence of the constructive solution of fine bubble generators on the concentration of oxygen dissolved in water. Adv. Mater. **2304**, 538–541 (2012). Online available since 2012/Jun/14 at www.scientific.net (2012) Trans Tech Publications, Switzerland
7. Băran, N., Căluşaru, I.M., Mateescu, G.: Influence of the architecture of fine bubble generators on the variation of the concentration of oxygen dissolved in water. Buletinul Stiintific al Universitatii Politehnica din Bucuresti, Editura Politehnica Press, serie D, Inginerie Mecanica **75**(3), 225–236 (2013)
8. Dobrovicescu, Al., Baran, N., si col.: Bazele termodinamicii tehnice, vol I. Elemente de termodinamica tehnica, Editura Politehnica Press, Bucuresti (2009)
9. Idelcik, I.E.: Indrumator Pentru Calculul Rezistentelor Hidraulice. Ed. Tehnica, Bucuresti (1984)
10. Kiselev, P.G.: Indreptar Pentru Calcule Hidraulice. Editura Tehnica, Bucuresti (1988)
11. Ionascu, G., Comeagă, C.D., Bogatu, L., Sandu, A., Besnea, D.: Modelling of material properties for MEMS structures. J. Optoelectron. Adv. Mater. **13**, 998 (2011)
12. Căluşaru, I.M., Baran, N., Pătulea, A.: Determination of dissolved oxygen concentration in stationary water. Rev. Chim. **63**, 1312 (2012)
13. Mateescu, G., Marinescu, A., Băran, N.: A new Constructing Fine Bubbles Generators. Bulletin of the Transylvania University of Brasov, vol. 2, p. 359. Brasov, Romania (2009)
14. Băran, N., Tănase, B., Rasha, M., Căluşaru, I.M.: Researches regarding the reduction of the water oxygenation time. Termotehnica **2**, 100 (2013)

Researches on the Measurement of the Dissolved Oxygen Concentration in Stationary Waters

Gheorghe I. Gheorghe[1(✉)], Octavian Donțu[2], Nicolae Băran[2], Ioana Corina Moga[3], Mihaela Constantin[2], and Eugen Tămășanu[2]

[1] National Institute of Research and Development in Mechatronics and Measurement Technique, Pantelimon Road, no. 6-8, District 2, Bucharest, Romania
incdmtm@incdmtm.ro

[2] Politehnica University of Bucharest, Splaiul Independenței no. 313, sector 6, Bucharest, Romania
octavdontu@yahoo.com, n_baran_fimm@yahoo.com, i.mihaelaconstantin@gmail.com, inregistrare@gmail.com

[3] DFR Systems SRL, Drumul Taberei no. 46, sector 6, Bucharest, Romania
corinamoga@yahoo.com

Abstract. The paper presents methods for measuring the dissolved oxygen concentration in water; the electrical method is presented by presenting an experimental installation designed and built in the laboratories of POLITEHNICA University in Bucharest. For the creation of fine air bubbles, a bubble generator with 0.1 mm orifices processed by spark-erosion was used. The results of the experimental researches on the measurement of dissolved oxygen in water are presented.

Keywords: Water aeration · Fine bubble generators · Spark-erosion

1 Introduction

The main purpose of water aeration, irrespective of the industry and the reason it is used, is to increase or maintain an optimal level of dissolved oxygen in water mass.

The oxygen required for the aeration process is taken from the atmospheric air and introduced into the water. In order for this aeration to be effective, uniform air dispersal must be ensured throughout the mass of water in a tank or basin; the air must be spread evenly so as to ensure the oxygen demand.

Dissolved oxygen content is the most important indicator of water quality.

Fishes, for example, need to survive a dissolved oxygen concentration up to 5 mg/dm^3 [1].

The amount of oxygen in the water is consumed by different biological or chemical processes. The amount of water remaining in these processes depends on the rate of deoxygenating and oxygenation rate (aeration), which can occur naturally or artificially [2, 3].

By aerating the water means transferring oxygen from atmospheric air into water, which is actually a phenomenon of transferring a gas into a liquid.

The most common method of removing impurities of organic nature under the action of a biomass of aerobic bacteria is the introduction of gaseous oxygen into the waste water.

G. I. Gheorghe (Ed.): ICOMECYME 2018, LNNS 48, pp. 29–40, 2019.
https://doi.org/10.1007/978-3-319-96358-7_4

Oxygen originates most frequently from atmospheric air, in this case the process known as water aeration (Fig. 1).

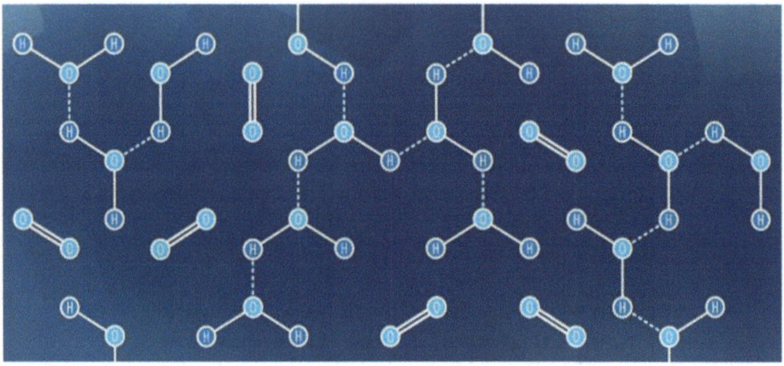

Fig. 1. Free oxygen molecules in water [4]

The dissolved oxygen in water is measured in mg O_2/dm^3.

From the above figure one can observe that each water molecule consists of an oxygen molecule connected to two hydrogen molecules. The oxygen molecules constituting the dissolved oxygen can be found among water molecules. The maximum amount of oxygen that can be dissolved in water depends on a number of physical and chemical parameters such as: atmospheric pressure, water temperature, salinity, turbulence of water [5, 6].

Water temperature is an important factor, so the warmer the water, the lower the dissolved oxygen concentration. So [1]:

- at t = 10 °C, in fresh, clean water, an amount of 11.3 mgO_2/dm^3 can be absorbed;
- at t = 25 °C, in clean water, only 8.3 mgO_2/dm^3 can be absorbed.

Oxygenation processes are found in the following areas [7, 8]:

- in sewage treatment and treatment plants;
- in disinfection (by ozonation) stations of raw water taken from a source in order to make it drinkable;
- in the chemical industry;
- in the food industry, fish industry, etc.;
- in water treatment and purification processes, oxygenation (sometimes referred to as aeration) is a basic process in ensuring water quality.

2 Presentation of Methods that Are Used to Measure the Dissolved Oxygen Concentration in Water

The analytical techniques used to determine the dissolved oxygen concentration in water are based on the following methods [9–11]:

2.1 Chemical Methods

The technique uses the IODOMETRIC method for determining the (DO) content in drinking water based on the Winkler process [12]. This method is increasingly less used because electrical and optical methods have emerged.

The main disadvantages of chemical methods are:

- the presence of catalysts used for deoxygenating;
- the presence of oxygen in reagents;
- the occurrence of errors in the titration process;
- cannot monitor DO instantly or continuously;
- it consumes more time than electrical or optical methods.

2.2 Electrical Methods

The electrical methods named in some works electrochemical methods are based on two techniques for measuring the dissolved oxygen concentration in water:

(a) galvanic technique where there is very little electrical voltage between electrodes, no external voltage is required [9];
(b) the technique of the polarographic process, in which an electric voltage (direct current) is applied between the two electrodes (cathode and anode).

In the following, only the polarographic process is analyzed.

The devices used to measure DO in water are called oxygen meters.

Overall, an oxygen meter is composed (Fig. 2) of a microprocessor (1) connected to a probe (3) which is introduced into water whose OD content has to be measured.

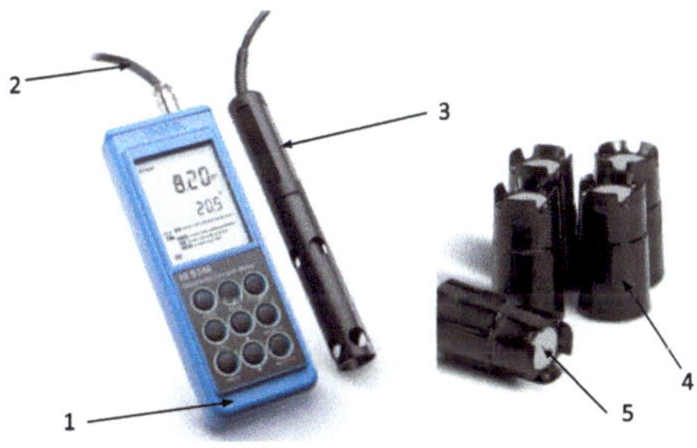

Fig. 2. Oxygen meter used for measurements. 1 - microprocessor; 2 - connecting cable; 3 - probe body; 4 - small cylinder containing an electrolyte solution; 5 - oxygen permeable Teflon membrane

The oxygen flowing through the permeable oxygen Teflon membrane (5) causes a change in the electric current between the cathode and the anode in a small cylinder containing an electrolyte solution (4); this change is proportional to the amount of oxygen that penetrated through the membrane and is displayed on the microprocessor screen in [mg/dm^3].

Disadvantages of Electrical Methods:

1. difficulty in calibrating and maintaining the device (electrode cleaning and membranes replacement);
2. electrolyte consumption during use of the device;
3. the devices are relatively expensive between 600–1500 €;
4. the disappearance of oxygen molecules in the vicinity of the sensor can only be prevented by keeping the sample of water or the probe in motion.

Points 1 and 2 can only be solved by regular calibrations and electrolyte changes.
Advantages of electric methods:

1. the apparatus is portable; measurements can be made in the laboratory, swimming pools, lakes, ponds;
2. the device monitors the DO concentration instantly or continuously;
3. no sampling equipment is required.

In the laboratory of the Department of Thermotechnics, Engines, Thermal and Refrigeration Equipment's there is such an oxygen meter that is used in experimental researches on water oxygenation.

2.3 Optical Methods

Optical methods include reflection spectroscopy, light analysis transmitted by translucent electrodes and ellipsometry.

In 2003, HACH LANGE became the first manufacturer of measuring instruments to launch the L.D.O. (Luminescent Dissolved Oxygen - Luminescent dissolved oxygen) to determine dissolved oxygen in water.

The device consists of a microprocessor (1), a connecting cable (2) and a probe inserted into the water (3) (Fig. 3).

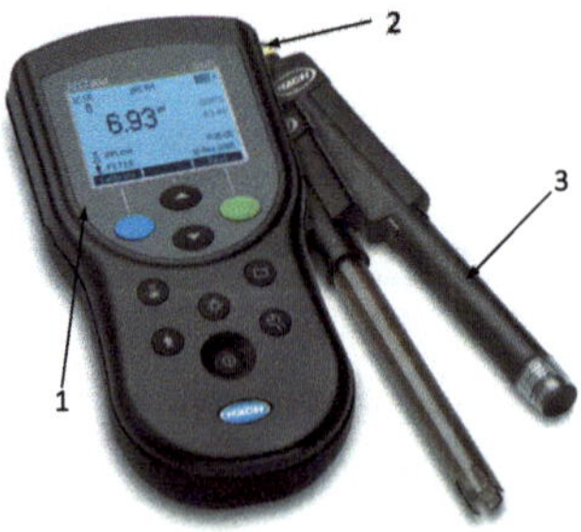

Fig. 3. DO meter based on L.D.O. 1 - microprocessor; 2 - connecting cable; 3 - probe that is inserted into the water

Technical data on the meter based on L.D.O. are presented in the table below:

Method of measurement	Luminescence, optical
Excitation	Pulses of blue light
Calibration	Not required
Measuring range	0.1 to 20 mg/dm^3 (ppm) O_2; 1–200% O_2 saturation; 0.1–50 °C
Accuracy	±0.1 mg/dm^3 $O_2 < 1$ mg/dm^3; ±0.2 mg/dm^3 $O_2 > 1$ mg/dm^3
Reproducibility	±0.5% of the final value of the measurement range
Response Time	T90 < 40 s (20 °C), T95 < 60 s (20 °C)
Temperature range	0–50 °C

The operating principle of L.D.O. is based on the physical phenomenon of luminescence; it is defined as the property of materials to emit light when excited.

For a suitable combination between luminophore and an excitation light of a given wave, the luminosity intensity and the time to disappear is dependent on the oxygen concentration around the luminophore.

The probe Hach Lange - L.D.O. is composed of two elements (Fig. 3):

(a) the head of the probe, on which there is a luminophore layer deposited on a transparent transfer material; the measurements of the head are screwed into the body of the probe that floats in water (Fig. 4);

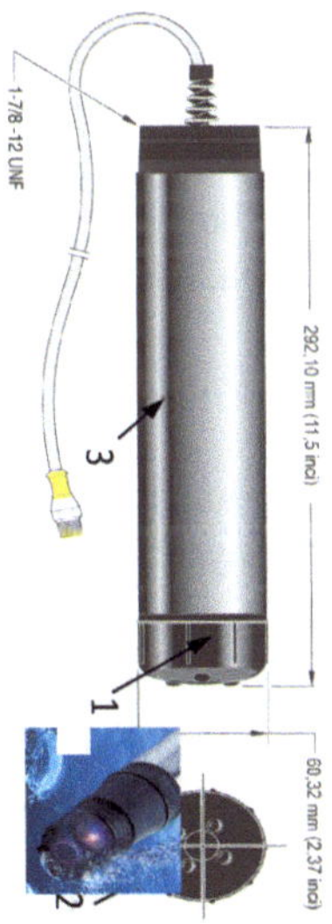

Fig. 4. HACH LANGE probe - L.D.O. 1 - probe head; 2 - layer luminophore; 3 - probe body

(b) the probe body, comprising a blue LED that emits the light needed to create luminescence, a red LED as a reference item and a photodiode.

In operation, the blue light emits pulses of blue light (excitation light) that reaches the luminophore, to which it transfers some of its radiant energy. As a result, some of the electrons in the luminophore layer leap from their original level to a higher energy level.

After a very short time, they return to their original level, emitting energy that they lose as red light (Fig. 5).

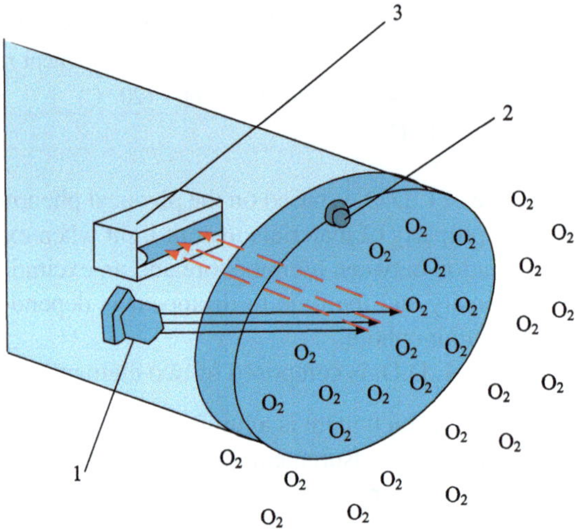

Fig. 5. Operating principle for HACH LANGE L.D.O. 1 - Blue LED; 2 - Red LED; 3 - photodiode

Oxygen molecules are able to absorb the energy of high-level electrons and allow them to return to their normal level without emitting red light.

The higher the oxygen concentration, the greater is the reduction in the intensity of the red light emitted.

As electrons return to low energy levels faster, the lifetime of the red light emitted is shortened.

To determine the oxygen concentration, the lifetime of the red light is evaluated. So, the DO measurement is based on the physical measurement of time.

Advantages of the L.D.O.:

(1) optical method L.D.O. measures the concentration of dissolved oxygen in water based on the measurement of an exact time.
(2) the method requires only that the oxygen molecules be in contact with the luminophor.
(3) any cleaning of the luminophore in the probe head does not affect the lifetime of red light emitted, which depends only on the oxygen concentration in the sample.
(4) all optical components of the probe are adjusted before each measurement using the red reference LED.

Before each measurement, it transmits a beam of light that reflects in the luminophore and passes through the entire optical system in the same way as light from luminescence.

Disadvantages of the L.D.O.:

(1) cleaning the probe head;
(2) changing the probe head once every 2 years.

2.4 Non-invasive Method

The current variety of applications, industrial or laboratory, requiring real-time monitoring of fluids variation in oxygen, has led to the development of several measurement methods.

Non-invasive measurement of dissolved oxygen concentration is the most recent method used in the food and beverage industry.

The determinations are accurate and can be done by means of a sensor applied to a transparent surface (glass or transparent plastic) (Fig. 6).

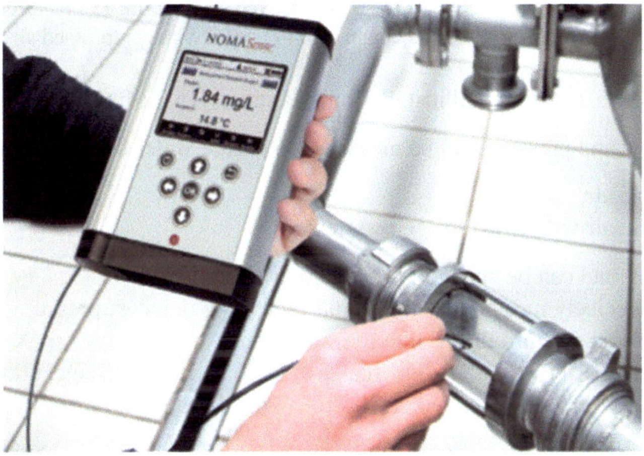

Fig. 6. Non-invasive device for measuring dissolved oxygen concentration

The principle of the measuring devices is that of oxo-luminescence [13, 14].

Figure 6 shows how to use a non-invasive device for measuring the concentration of dissolved oxygen in water passing through a pipe [14].

The main features of this device are: it uses a non-invasive, non-destructive method; applicability in gaseous or liquid media; long life of sensors without complicated calibration or maintenance operations; usable in industrial or laboratory environments; easy to use, portable and versatile; accurately determine the dissolved oxygen content in water.

3 Experimental Researches on the Determination of Dissolved Oxygen in Water by the Electrical Method

The fine bubble generator (FBG) is fed by its two ends, namely through the 16 and 17 pipes (Fig. 7).

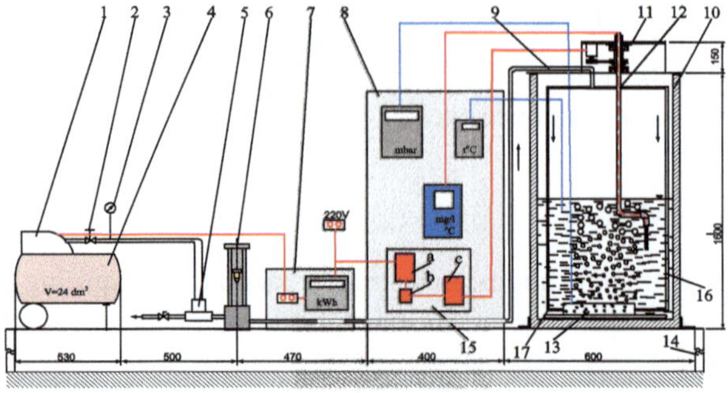

Fig. 7. The scheme of experimental installation for research on water oxygenation. 1 - compressed air compressor; 2 - pressure reducer; 3 - manometer; 4 - compressed air tank V = 24 dm^3; 5 - T-joint; 6 - rotameter; 7 - electric panel; 8 - panel with measuring devices; 9 - compressed air pipeline to FBG; 10 - water tank; 11 - probe actuation mechanism; 12 - Oxygen probe; 13 - FBG of rectangular shape; 14 - support for the installation; 15 - electronics control: a - power supply, b - switch, c - control element; 16, 17 - FBG with compressed air

Regarding the experimental installation it is stated [15, 16]:

– the gas flow and pressure can be measured to ensure water oxygenation:

- atmospheric air,
- atmospheric air with low-nitrogen content.

– hydrostatic load can be changed in range 0.5; 1.0; 1.5 m.
– it can be precisely measured with digital indicating devices: pressure, temperature and flow of the gas introduced into the water tank.
– one can measure the instantaneous change of dissolved oxygen concentration in water or at time Δt with the oxygen meter whose probe is actuated by an electric mechanism.

For the operation of the experimental installation, electric current is needed to drive the electro compressor, the mechanism of rotating the oxygen sensor probe into the water tank.

The system works as follows: the air compressed by an electro compressor 1 (Fig. 7) accumulates at p = 1.5–2 bar in a 24 dm^3 tank. Subsequently, the air passes through the reducer (2) through the manometer (3) and reaches the FBG (13).

During an experiment, the volumetric air flow rate, the pressure at the entrance to the FBG (p_{FBG} and hydrostatic load (H) are maintained constant.

The panel (15) with the control electronics provides, via the mechanism (11), the rotation of the oxygen sensor probe in the water tank at a speed of 0.3 m/s.

For measuring the pressure and temperature of the air and the dissolved oxygen concentration in the water, digital indicating devices are provided on the panel (8) of Fig. 7, namely:

(a) for the flow rate measurement a rotameter with a scale of 0–2200 dm^3/h for air was provided.

(b) a manometer with a digital indication in the range 0–190 mbar was provided for the pressure measurement.
(c) the measurement of temperature was performed with a digital thermometer with a scale of 50–150 °C.
(d) the dissolved oxygen concentration in water was measured on the basis of the electric method [17, 18] using a polarographic probe oxygen sensor.

To carry out the measurements, the oxygen sensor was rotated in the water tank with 2 rot/s; for its rotation an electro-mechanical mechanism realized in the Department of Thermotechnics, Engines, Heat and Refrigeration Equipment's was designed.

The measurement of the dissolved oxygen concentration in water is based on the electrical method. The oxygen meter has a polarographic probe to be displaced during the measurements; the displacement consists of a rotation movement of 0.3 m/s (value required in the Oxygen Leaflet). The probe radius is 0.125 m.

The speed that the mechanism ensures is thus established [19, 20]:

$$w = \omega \cdot r \left[\mathrm{m/s}\right] \tag{1}$$

$$\omega = \frac{w}{r} = \frac{0.3}{0.125} = 2.4 \left[\frac{\mathrm{rad}}{\mathrm{s}}\right] \tag{2}$$

$$\omega = \frac{2\pi \cdot n}{60} \tag{3}$$

$$n = \frac{60 \cdot \omega}{2\pi} = \frac{60 \cdot 2.4}{2\pi} = 22.92\,\mathrm{rot/min} \tag{4}$$

The purpose of the researches is to experimentally determine the variation in dissolved oxygen concentration in water.

Figure 8 shows the FBG with 152 orifices Ø 0.1 mm in operation [20–22].

Fig. 8. FBG with 152 orifices Ø 0.1 mm in operation

For experimental researches, as initial data it is specified:

- the airflow introduced in the FBG: 600 dm^3/h
- the air pressure at the FBG entrance: p = 573 mm H_2O
- water temperature: t = 24 °C
- the initial concentration value: $C_0 = 5.48$ mg/dm^3
- Corresponding to the water the air pressure at the FBG entrance: p = 573 mm H_2O temperature, the value of the saturation concentration: Cs = 8.4 mg/dm^3

The experimental researches resulted are presented in Table 1. OD-Oxygen Dissolved in Water.

Table 1. Experimental data obtained from the FBG of rectangular shape with 152 orifices

No	0	1	2	3	4	5	6	7	8
τ [min]	0	15	30	45	60	75	90	105	120
t_{H2O} [°C]	23.7								
t_{aer} [°C]	24.1								
OD [mg/dm^3]	5.4	7.3	7.85	8.0	8.2	8.3	8.31	8.37	8.39

Based on the data in Table 1, the function C = f (τ) of Fig. 9 was plotted.

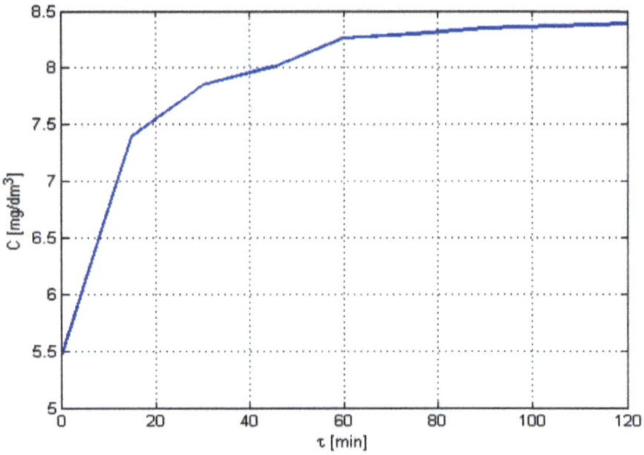

Fig. 9. The variation of the dissolved oxygen concentration in water, in time C = f (τ)

The presented experimental results coincide very well with the theoretical results of the function C = f (τ) presented in the papers [9, 18, 23, 24].

4 Conclusions

The measurement of the dissolved oxygen concentration in water is necessary because it ensures the existence of water life forms within certain limits.

Out the methods of measuring the dissolved oxygen concentration in water, the most commonly used is the electrical method; it has the advantage of instantly indicating the change in oxygen concentration on a computer screen. It has the disadvantage that during the measurement, the probe must be driven in rotation by an electro-mechanical mechanism.

The non-invasive method presented in Sect. 2.4 is a modern method that makes it easy and accurate to perform measurements.

Acknowledgements. This work was supported by a grant of the Romanian National Authority for Scientific Research and Innovation, CCCDI – UEFISCDI, project number Manunet – MNET17/ENER2307 – CEBIOTREAT, within PNCDI III.

References

1. Oprina, G., Pincovschi, I., Băran, G.: Hidro-Gazo-Dinamica Sistemelor de aerare echipate cu generatoare de bule. POLITEHNICA PRES, Bucureşti (2009)
2. Robescu, D., Robescu, D.L.: Procedee, instalaţii şi echipamente pentru epurarea apelor, Litografia UPB, Bucureşti (1996)
3. Droste, L.R.: Theory and Practice of Water and Wastewater Treatment. Wiley, New York (1996). ISBN 978-0-471-12444-3
4. Căluşaru, I.: Influenţa proprietăţilor fizice ale lichidului asupra eficienţei proceselor de oxigenare, Teză de doctorat, Universitatea Politehnica din Bucureşti, Facultatea de Inginerie Mecanică şi Mecatronică (2014)
5. Pătulea, A.S.: Influenţa parametrilor funcţionali şi a arhitecturii generatoarelor de bule fine asupra eficienţei instalaţiilor de aerare. Teză de doctorat, Universitatea Politehnica din Bucureşti (2012)
6. Oprina, G.: Contribuţii la hidro-gazo-dinamica difuzoarelor poroase, Teză de doctorat, Universitatea Politehnica din Bucureşti, Facultatea de Energetică (2007)
7. Robescu, D., Robescu, D.L., Verestoy, A.: Fiabilitatea proceselor, instalaţiilor şi echipamentelor de tratare şi epurare a apelor. Tehnică, Bucureşti (2002)
8. Robescu, D.L., Lanyi, S., Verestoy, A., Robescu, D.: Modelarea şi simularea proceselor de epurare. Tehnică Bucureşti (2004)
9. Băran, N., Patulea, A.S., Căluşaru, I.M.: The determination of the oxygen transfer speed in water in nonstationary conditions. In: International Proceedings of Computer Science and Information Technology, Mechanical Engineering, Robotics and Aerospace, pp. 267–272. IACSIT Press (2011). ISSN 2010-460X, ISBN 978-981-07-0420-9
10. Pătulea, A., Băran, N., Căluşaru, I.M.: Measurements of dissolved oxygen concentration in stationary water. World Environ. **2**(4), 106–109 (2012)
11. Căluşaru, I.M., Costache, A., Băran, N., Ionescu, G.L., Donţu, O.: The determination of dissolved oxygen concentration in stationary water. Appl. Mech. Mater. **436**, 233–237 (2012). (Trans Tech Publications, Switzerland)
12. Pincovschi, I.: Hidrodinamica sistemelor disperse gaz-lichid, Teză de doctorat, Universitatea Politehnica din Bucureşti (1999)
13. Mitchel, T.O.: Luminescence based measurement of dissolved oxygen in natural waters (2006). www.hachenvironmental.com
14. Noninvasive, D.O.: Measurement http://www.nomacerc.com/enology/nomasense_02_p300
15. Căluşaru, I., Băran, N., Pătulea, A.: The influence of the constructive solution of fine bubble generators on the concentration of oxygen dissolved in water. Adv. Mater. Res. **538–541**, 2304–2310 (2012). (Trans Tech Publications, Switzerland)

16. Pătulea, A., Căluşaru, I.M., Băran, N.: Researches regarding the measurements of the dissolved concentration in water. Adv. Mater. Res. **550–553**, 3388–3394 (2012). (Trans Tech Publications, Switzerland)
17. Căluşaru-Constantin, M., Tănase, E.B., Băran, N., Mlisan-Cusma, Rasha: Researches regarding the modification of dissolved oxygen concentration in water. IJISET - Int. J. Innov. Sci. Eng. Technol. **1**(6), 228–231 (2014)
18. Constantin, M., Băran, N., Tănase, B.: A new solution for water oxygenation. Int. J. Innov. Res. Adv. Eng. (IJIRAE) **2**(7), 49–52 (2015)
19. Dobrovicescu, A., Băran, N., Chisacof, A.: Bazele termodinamici Tehnice, Elemente de Termodinamică Tehnică. POLITEHNICA PRESS (2009)
20. Băran, Gh., Pincovschi, I., Oprina, G., Bunea, F.: Performanţe ale generatoarelor de bule fine. Revista Hidrotehnica, Bucureşti **53**(3–4), 27–32 (2008)
21. Tănase, B., Besnea, D., Mlisan, R., Constantin, M., Băran, N.: Constructive solutions for the achievement of fine bubble generators based on micro-drilling technologies. IJISET - Int. J. Innov. Sci. Eng. Technol. **2**(2), 46–50 (2015)
22. Tănase, E.B., Băran, N., Mlisan, R.: An efficient solution for water oxygenation. Asian Eng. Rev. **1**(3), 36–40 (2014)
23. Căluşaru, I.M., Băran, N., Pătulea, A.: The influence of the constructive solution of fine bubble generators on the concentration of oxygen dissolved in water. Adv. Mater. Res. **538–541**, 2304–2310 (2012). (Trans Tech Publications, Switzerland)
24. Căluşaru, I.M., Băran, N., Pătulea, A.: Researches regarding the transfer of oxygen in water. In: The 3rd International Conference on Mechanic Automation and Control Engineering (MACE 2012), pp. 2617–2620 (2012). July 27th–29th, published by IEEE Computer Society CPS, and then submitted to be indexed by Ei Compendex, Baotou, China

Experimental Results of Turbo-Aggregate Vibroacoustic Diagnosis Obtained with Vibro-Expert System for One Turbo Aggregate in Lukoil Refinery

Cornel Marin(✉) and Ionel Rusa

Valahia University of Târgovişte, FIMM, Str. Aleea Sinaia, No. 13, Targoviste, Romania
marin_cor@yahoo.com, ionelrusa@gmail.com

Abstract. Proactive maintenance is a relatively new concept used today by the exploitation of manufacturing and energy production systems consisting of vibro acoustic monitoring of installations and equipment with professional equipment, such as Expret-Vibro and PROFISIGNAL software. These vibration level measurements are required for vibro acoustic diagnosis and for timely programming of repairs that are being challenged before accidental malfunctions occur. Electric steam turbines are used to produce electrical energy in refineries, which are particularly complex and are equipped with a turbine shaft located on several sliding or rolling bearings. Vibration sensors are mounted on bearing housings and are formed from one-axial, biaxial and three-axial accelerometers for absolute magnitudes (P-P and RMS speeds) or laser proximal sensors (relative and offset displacements). These sensors transmit the signals of data acquisition cards and amplifiers for data processing.

Keywords: Vibration diagnosis · Vibro-expert system · Proactive maintenance

1 Introduction

Measurement of the vibration level was performed on the TA4 Turbo Aggregate in the LUKOIL operating state at normal mode parameters. Diagnosis of the operating state is based on measurements of the absolute vibrations in the TA4 turbine units by means of one-axial accelerometers VIBRASENS 101.51-9 and measurements of the relative vibrations of the spindle (orbital) by means of some proximity sensors ROLS-W - produced by Monarch Instruments.

The four bearings on which the sensors are mounted are shown in Fig. 1. The normal operating parameters of the LUKOIL TA4 torque-aggregate are:

- Power: 32 MW;
- Speed: 3000 rpm;
- Stator type: 11 kV, 2100 A,
- Tip Rotor: 532 A
- Water flow: 150 mc/h

G. I. Gheorghe (Ed.): ICOMECYME 2018, LNNS 48, pp. 41–55, 2019.
https://doi.org/10.1007/978-3-319-96358-7_5

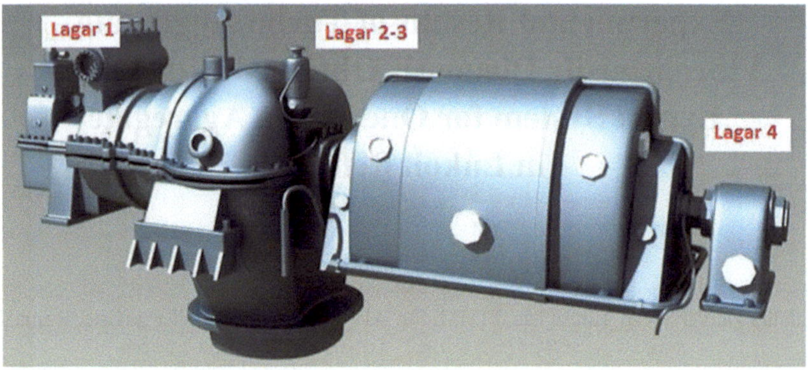

Fig. 1. 3D Model virtual of Turbo Aggregate TA4

2 Expert Vibro Data Acquisition and Processing System

The Expert Vibro Diagnostic System Integration Scheme and Modules in the April-May 2017 measurements on the TA4 Turbo Aggregate in the normal operation of the LUKOIL frame is shown in Figs. 2 and 3 includes the following equipment and modules:

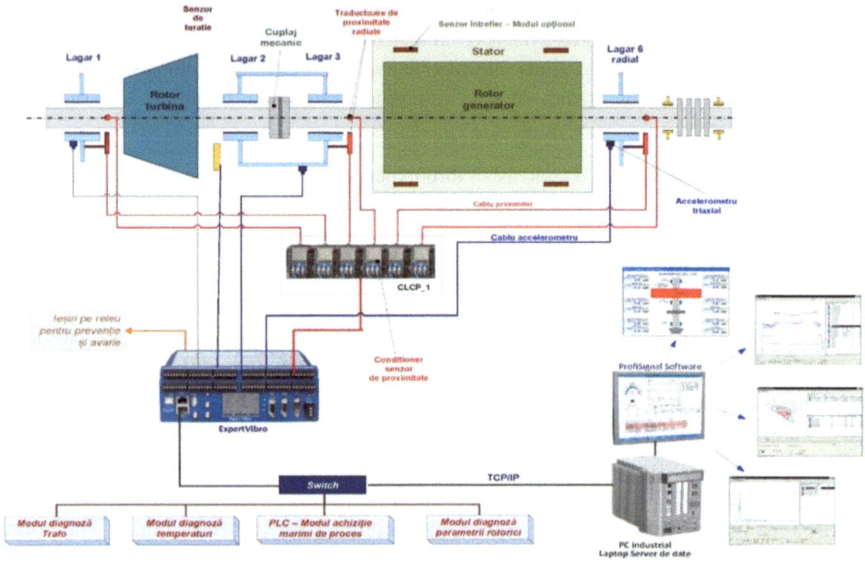

Fig. 2. Expert Vibro diagnostic system integration block diagram

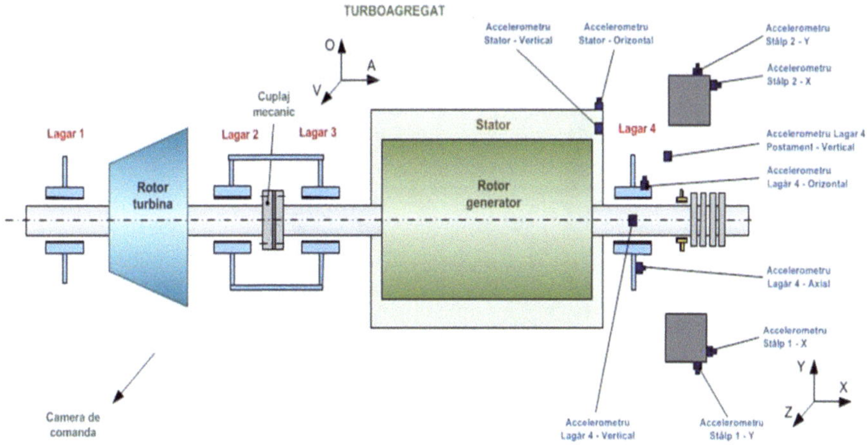

Fig. 3. Location of vibration measurement points on bearing 4 (Axial, Vertical and Horizontal), pillars (X and Y direction) and base (vertical)

- Expert Vibro Diagnostic System, PC Computer
- Speed sensors, accelerometers and laser proximity transducers, air gap sensors
- The TRAFO diagnosis module
- The way it diagnoses temperatures
- Process acquisition process – PLC
- Module diagnosis Rotor parameters

3 Technical Conditions of Measurement, Operating Regimes and Types of Measurements

Technical conditions for vibration measurements and experimental data processing are in according with international standards: ISO 7919-1: 1996 [9]; ISO 7919-3: 1996 [10]; ISO 10816-1: 1995 [11] and ISO 10816-3: 2009 [12].

One axial and three axial accelerometers were used along the X, Y, Z directions as shown in the positioning scheme of the measuring points in Fig. 3, in particular, the positioning of the sensors on the bearing 4 in Figs. 4, 5 and 6, the two posts and the post of the bearing, according to the vertical directions - V, Horizontal - O and Axial - To the bearing and the post, respectively to the X, Y and Z directions for the pillars. Vibration measurements were performed for the following operating modes:

Fig. 4. Mounting of accelerometers on bearing

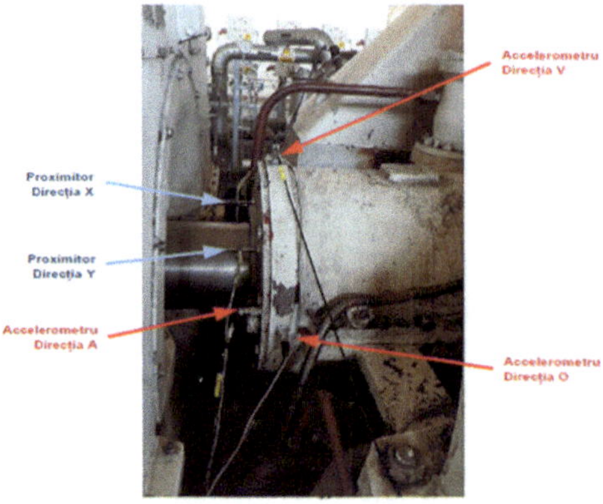

Fig. 5. Mounting of accelerometers and proximity sensors on bearings 2-3

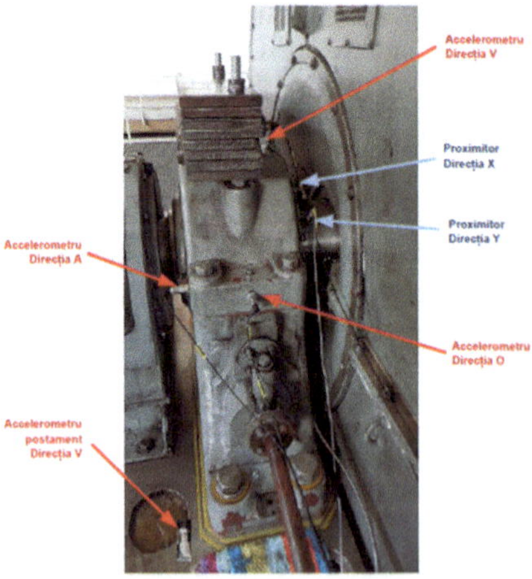

Fig. 6. Mounting of accelerometers and proximity sensors on bearings 4 and basis

a. At 3000 rpm and 20 MW power with 8 metal plates mounted on bearing 4
b. At 3000 rpm and the power of 14 MW with 8/7/6/5/4/3/1/0 metal plates mounted on bearing 4

With the Expert Vibro Diagnostic System, global vibration values were monitored:

- Relative vibrations of the Turbo Aggregate shaft TA4 according to ISO 7919-3
- Absolute vibrations of turbine aggregate TA4 according to ISO 10816-3
- Amplitude of base - component 1x - 50 Hz, 2x ,- 100 Hz, 3x 150 Hz recorded on Bearing 4 and its mount in vertical direction.
- Waveforms and Frequency Spectra for acceleration, velocity and displacement
- Orbit cams and vibrations absolute and relative etc.

ISO Standard 7919-3: 1996 and ISO 10816-3: 2009 establishes the performance ratings of the Turbo Aggregate when it is stable from the energy point of view, as shown in Table 1:

Table 1.

		ISO7919-3:1996	**ISO 10816-3:2009**	
Machine		Turboagregate 3000 rpm, P =32 [MW]	Turboagregate 3000 rpm, P =32 [MW]	
Measurement parameter		**Displacement Pick - pick [µm]**	**Speed RMS [mm/s]**	**Displacement Pick - pick [µm]**
Performance ratings	Good – A	under 85	under 2.3	under 29
	Usable – B	85–160	2.3–4.5	29–57
	Admitted under supervision – C	160–240	4.5–7.1	57–90
	Not allowed – D	over 240	over 7.1	over 90

- Rating A - Good/Vibration rating of newly installed machines enters this area.
- Rating B - Usable/Machine vibrations within this area are normally considered acceptable for unrestricted operation in the long regime operation.
- Qualification C - Admitted under supervision/Machines where the vibrations within this area are normally considered unsatisfactory for continuous, long-term operation.
- Qualification D - Not allowed/Machines where vibrations within this area are considered to be severe enough to cause damage to the machine.

4 Results Obtained Using Expert Vibro Data Acquisition, Measurement and Data Acquisition System

Following the vibration measurements, high vibration amplitudes relative to the bearing 2-3 were measured, and on the bearing 4 the values of the relative vibration amplitudes are small. The vibration amplitudes have been recorded for both working modes, the machine's performance rating is usable - according to the Global Relative Vibration Level Bulletin No. 1 [7].

From the diagnosis of the relative vibrations of the intermediate bearings 2-3 and the end bearing 4, there results a slight loading of the bearing 2-3 which may lead to the de-shafting of the shafts in the vertical direction according to Fig. A.9 - Orbital displacement bearing 2-3 and Fig. A.10 - Orbital Movement 4, Annex 1 [8].

From the diagnosis of the absolute vibrations of the intermediate bearings 2-3 and the end bearing 4, it is found that the machine qualifies as Good for all measuring regimes, according to Bulletin No. 3 [7].

From the diagnosis of absolute vibrations, the amplitudes of the RMS vibration velocity, these are classified as Admitted under Surveillance for both the 14 MW power regime and the 20 MW regime, according to the Bulletin No. 2 [7]. The highest RMS vibration amplitudes were recorded on the bearing 4 in the axial direction.

From the information provided by the beneficiary, it results that the bearing 4 resonates at the speed of 3000 rpm - 50 Hz. For this purpose a series of measurements of the overall vibration level at the bearing 4: 1x - 50 Hz, as well as the spectral components 2x - 100 Hz/3x - 150 Hz, were performed in different variants of dynamic absorbers that

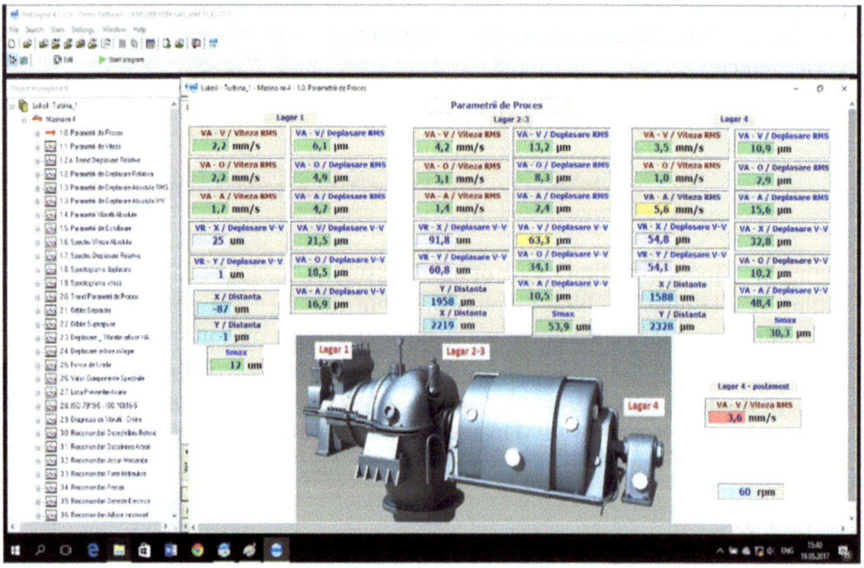

Fig. 7. Turbo aggregate Vibration Parameters - Measurement Mode: 5 plates on the bearing 4

were mounted on shaft 4: from 0 to 8 metal plates mounted in the upper part of the bearing (Fig. 7).

From the diagnosis of the absolute vibrations it results that the smallest vibrations were recorded on the bearing 4 on which 5 plates were mounted, obtaining the overall value of 5.57 mm/s, the component 1x of 4.59 mm/s, the 2x component of Fig. 2. Figure A.13 - Frequency Spectrum Bearing 4 - Axial/Horizontal/Vertical - Appendix 1 [8].

The maximum measured value RMS 5.57 mm/s exceeds the value prescribed by ISO 10816-3: 2009 of 4.5 mm/s for the Usable rating and falls under the Admitted under supervision rating.

5 Conclusions and Recommendations

For the correct diagnosis of the machine, it is recommended to carry out additional measurements in other working modes such as:

- Raising the rpm from 0 rpm to the nominal speed of 3000 rpm and over 3000 rpm to determine the displacement of the center position of the spindle in the bearing (orbital) and the resonance frequency for the bearing 4. The relative and absolute global vibration shall be measured, the relative displacement of the spindle, component 1x, 2x simultaneously with phase measurement.
- Nominal idle speed at rated rpm of 3000 rpm, at an elevated void and at a maximum load of 20 MW.
- Measuring the transmissibility of three-axial vibrations from the bearing 4 to the base/basis plate - the concrete foundation of the generator casing.

The first two types of measurements were not possible due to the unavailability of the machine. If the increased speed vibration is due to the resonance frequency of the bearing 4, it can be reduced by mounting masses on the bearing.

Appendix

See Figs. A.1, A.2, A.3, A.4, A.5, A.6, A.7, A.8, A.11, A.12, and A.14.

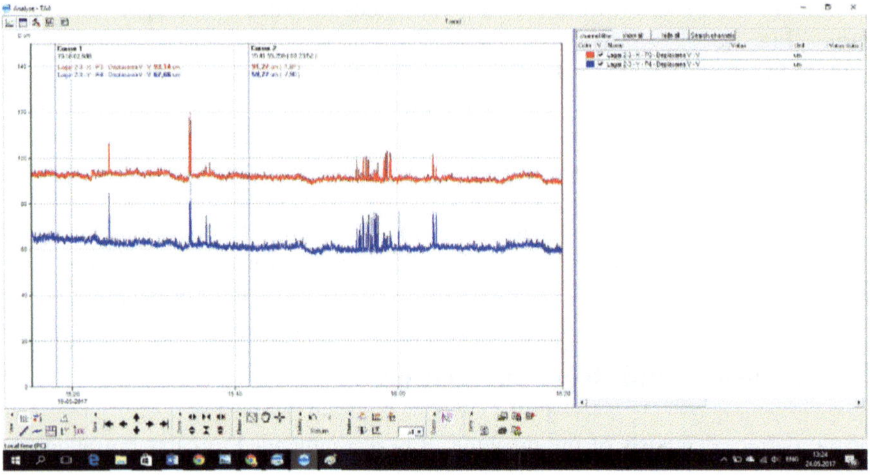

Fig. A.1. Trend vibration parameters relative to bearing 2-3. Measuring mode: 8 … 0 plates on bearing 4, cursor 1-8 plates/cursor 2-5 plates

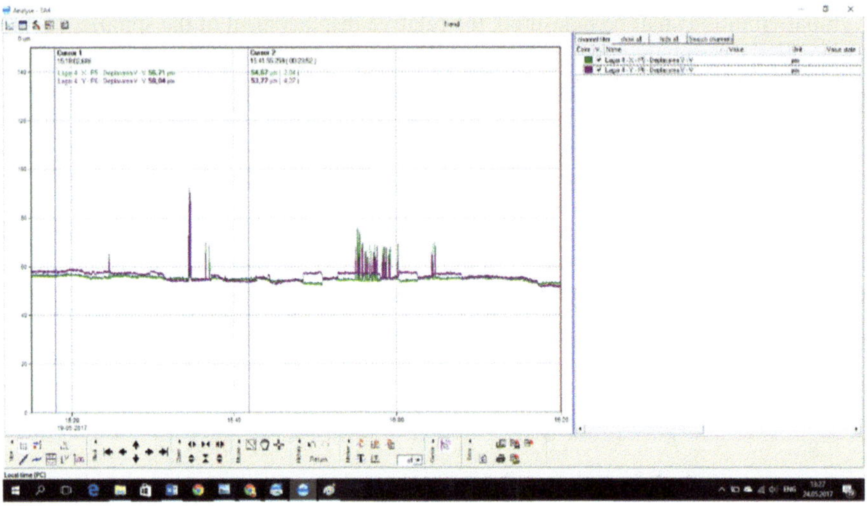

Fig. A.2. Trend vibration parameters relative to bearing 4. Measuring mode: 8 … 0 plates on bearing 4, cursor 1-8 plates/cursor 2-5 plates

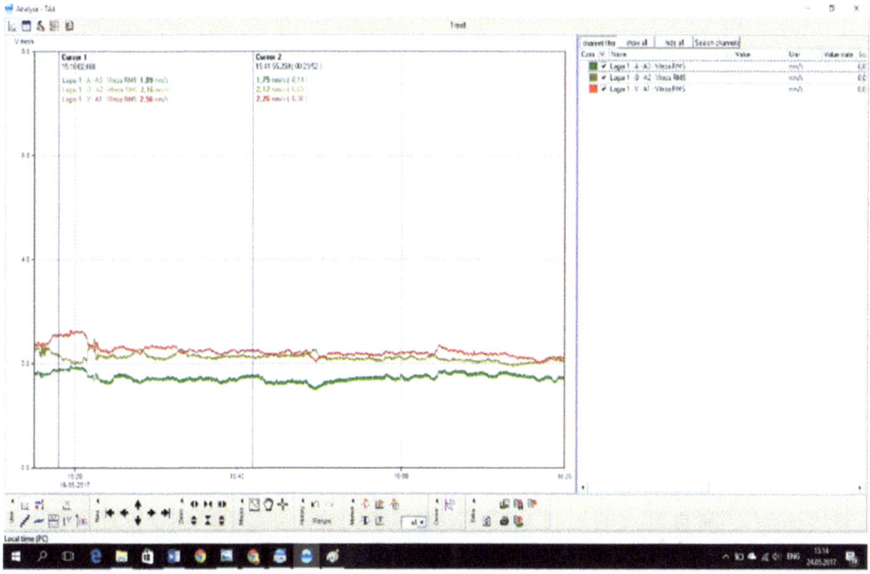

Fig. A.3. Trend vibration parameters absolute to bearing 1. Measuring mode: 8 … 0 plates on bearing 4, cursor 1-8 plates/cursor 2-2 plates

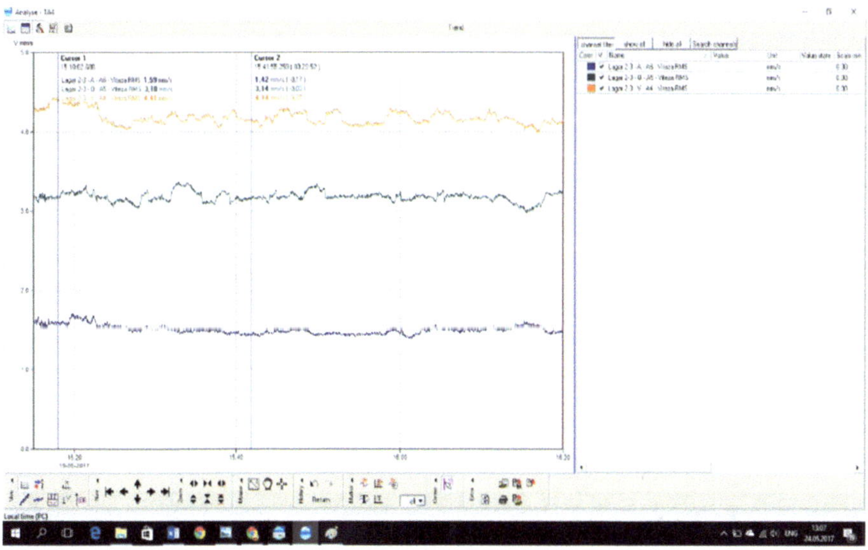

Fig. A.4. Trend vibration parameters absolute to bearing 2-3. Measuring mode: 8 … 0 plates on bearing 4, cursor 1-8 plates/cursor 2-2 plates

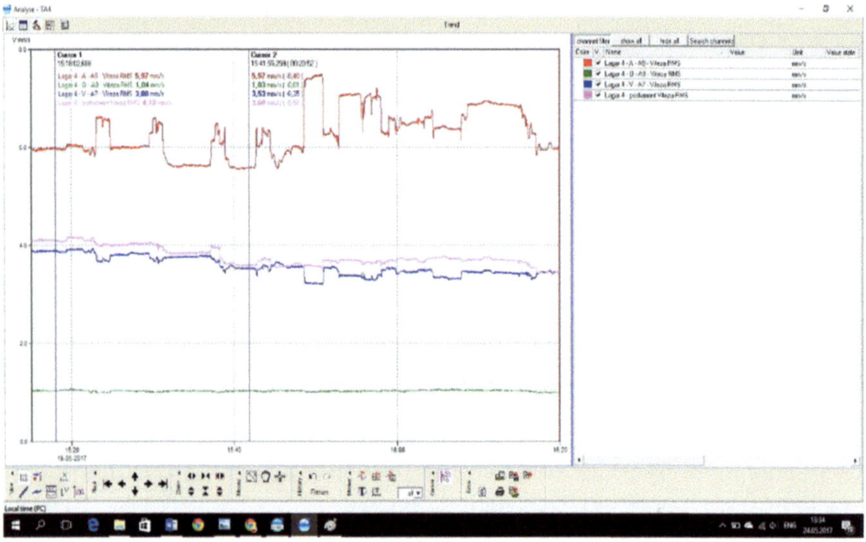

Fig. A.5. Trend absolute speed bearing 4 and base. Measuring mode: 8 … 0 Plates on bearing 4, Cursor 1-8 Plates/Cursor 2 - Plates

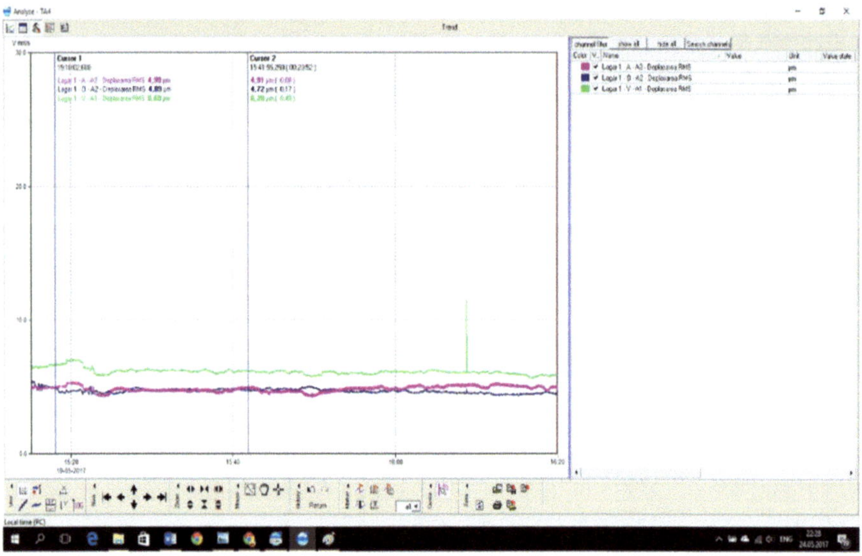

Fig. A.6. Trend absolute speed bearing 1 and base. Measuring mode: 8 … 0 Plates on bearing 4, Cursor 1-8 Plates/Cursor 2 – Plates

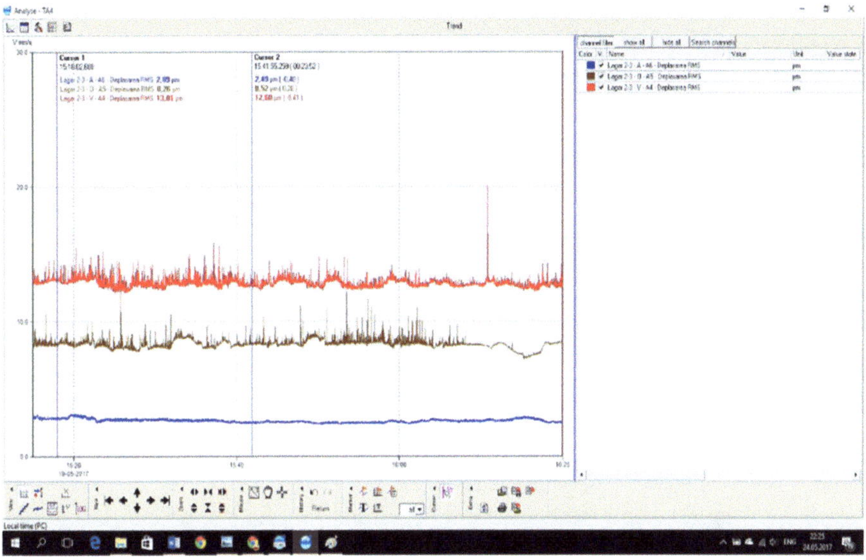

Fig. A.7. Trend absolute speed bearing 2-3 and base. Measuring mode: 8 … 0 Plates on bearing 4, Cursor 1-8 Plates/Cursor 2 – Plates

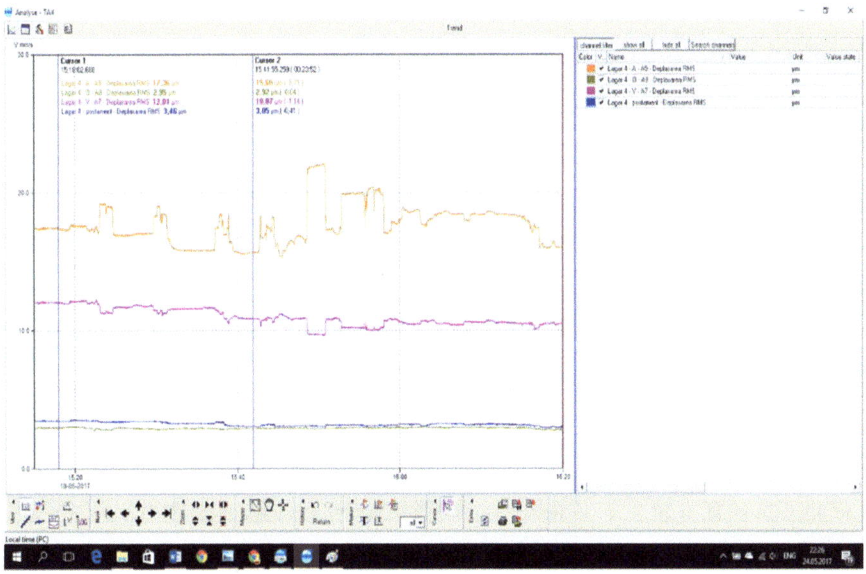

Fig. A.8. Trend absolute speed bearing 4 and base. Measuring mode: 8 … 0 Plates on bearing 4, Cursor 1-8 Plates/Cursor 2 – Plates

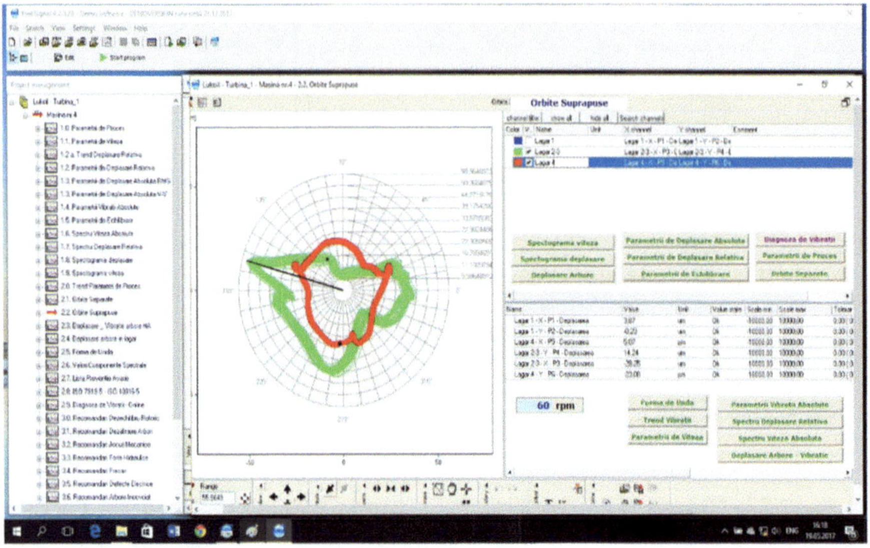

Fig. A.9. Orbital movement bearing 2-3 and bearing 4. Measuring regime: 5 plates on the camp 4

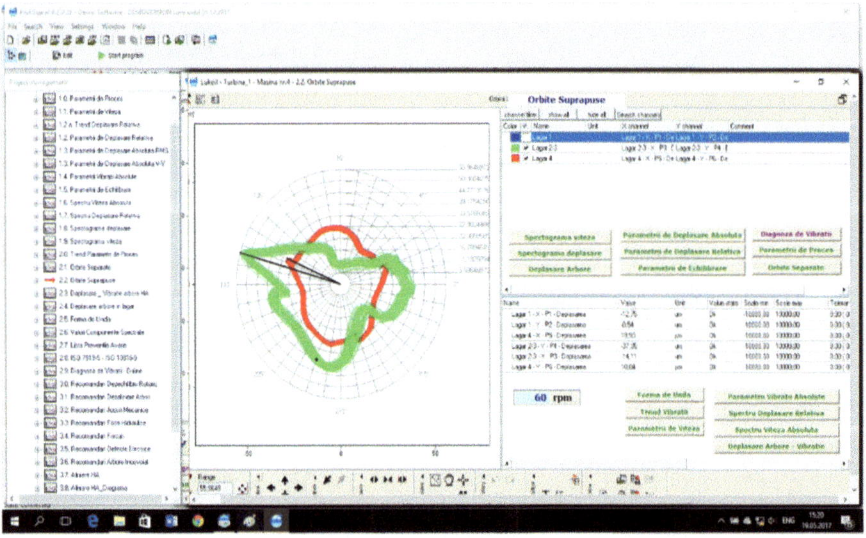

Fig. A.10. Orbital movement bearing 2-3 and bearing 4. Measuring regime: 8 plates on the camp 4

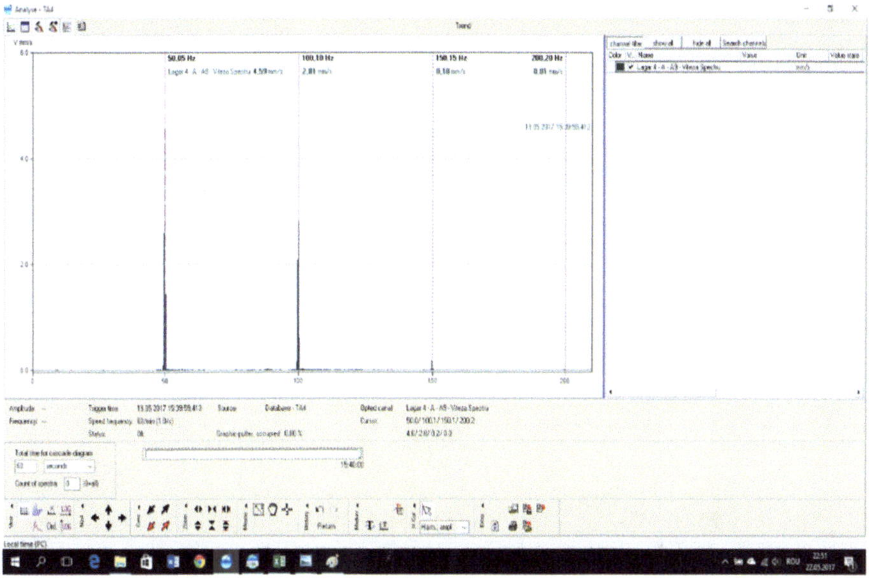

Fig. A.11. Frequency Spectrum Bearing 4 - Axial - Measuring mode: 5 plates on bearing 4

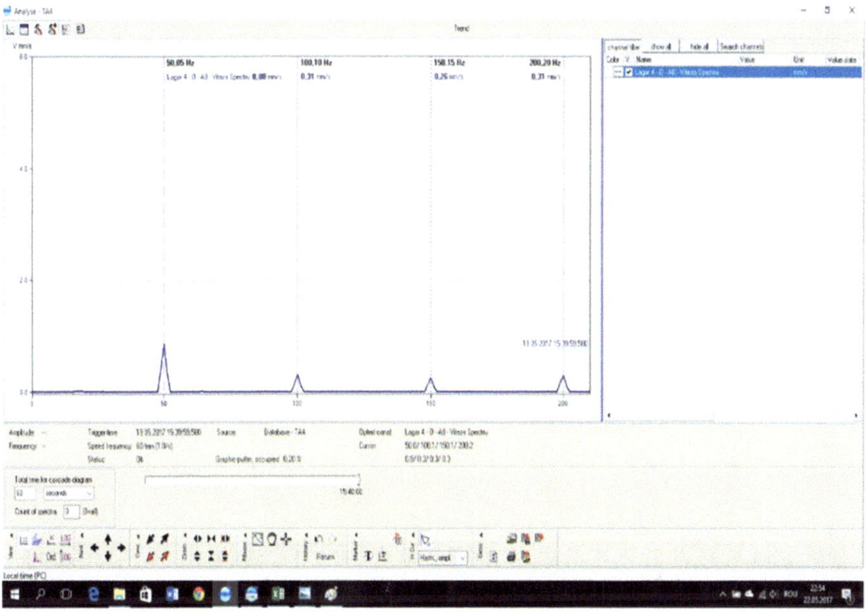

Fig. A.12. Frequency Spectrum Bearing 4 - Horizontal - Measuring mode: 5 plates on bearing 4

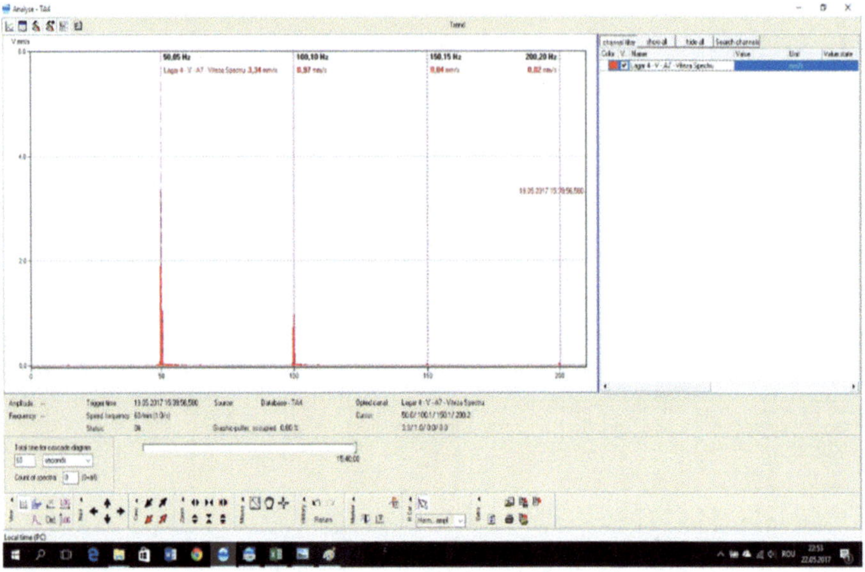

Fig. A.13. Frequency Spectrum Bearing 4 - Vertical - Measuring mode: 5 plates on bearing 4

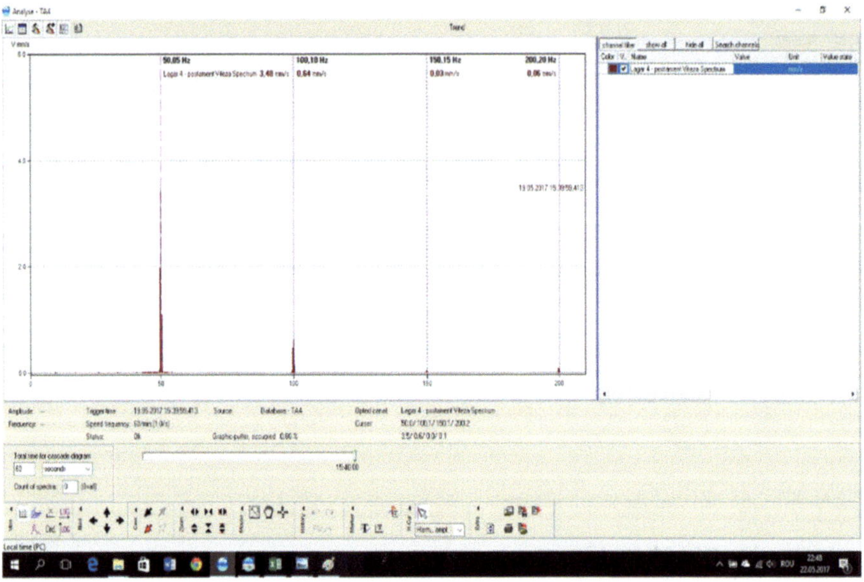

Fig. A.14. Frequency Spectrum Bearing 4 – Vertical Postament - Measuring mode: 5 plates on bearing 4

References

1. Introducere în vibrații - Ref.Doc.MI 119 – Notă tehnică / Mobil Industrial – Pitești (2013)
2. Mentenanța utilajelor dinamice vol. 1 – Mobil Industrial AG – Pitești (2011)
3. Vibration Calibration Technique and basics of Vibration Measurement - Torben R. Licht – Singapore 2011
4. An Introduction to Vibration Analysis Theory and Practice
5. Automated machinery maintenance – Bill Powel, Tony Burnet
6. Bill Duncan – "Bureau of Reclamation Plumb and Alignment Standards for Vertical Shaft Hydro-units"
7. S.C. VIBRO SYSTEM S.R.L. www.vibrosystem.ro – Buletin de măsurare a nivelului global de vibraţii relative NR. 1; Buletin de măsurare a nivelului global de vibraţii relative NR. 2; Buletin de măsurare a nivelului global de deplasare absolută NR. 3
8. S.C. VIBRO SYSTEM S.R.L. www.vibrosystem.ro – Raport tehnic privind măsurătorile de vibrații efectuate cu sistemul de diagnoză expert vibro la turboagregatul TA 4 – CET II LUKOIL ENERGY&GAS S.R.L. mai 2017
9. ISO 7919-1: 1996- Vibraţii mecanice ale maşinilor. rotativi şi riteriile de evaluare. Partea 1: Prescripţii generale
10. ISO 7919-3: 1996 Vibraţii mecanice. Evaluarea vibraţiilor maşinilor prin măsurători ale arborelui rotativ. Partea 3: Masini industriale cuplate
11. ISO 10816-1: 1995 Vibraţii mecanice. Evaluarea vibraţiilor maşinilor prin măsurători pe părţile non-rotative. Partea 1: Prescripţii generale
12. ISO 10816-3 - 2009 Vibraţii mecanice. Evaluarea vibraţiilor maşinilor prin măsurători pe părţile non-rotative. Partea 3: Maşini industriale cu puterea nominala peste 15 kW şi turaţia între 120 rpm și 15000 rpm măsurate in situ

Applications of Additive Technologies in the Food Industry

Daniel Besnea(✉), Octavian Dontu, Victor Constantin, Alina Spanu, Ciprian Rizescu, and Edgar Moraru

Politehnica University of Bucharest, Splaiul Independenţei nr. 313, sector 6, Bucharest, Romanía
d_bes@yahoo.com, octavdontu@yahoo.com, victor.f.constantin@gmail.com, spanu_alina@yahoo.com, ciprianrizescu2001@yahoo.com, eddy_milan91@yahoo.com

Abstract. The article presents a new concept that can be used in the food industry by manufacturing customized food products with 3D printers. This technique allows the production of personalized food products both as shape, colour, aroma and even as a nutritional value. Applications in the food industry can offer an engineering solution for new products and a tool for design specialists by creating food products with complex shapes that cannot be obtained by classical processes. Also, this technology can be a solution for researchers exploring cosmic space and long-distance space missions.

Keywords: Customized food · 3D food industry printers
Foods with complex forms

1 Introduction

Food manufacturers want to develop new techniques in this 3D printing field due to their unique features. Digital techniques used in gastronomy to produce customized food products that have the form, flavour, colour, texture and even nutritional value are strictly controlled because it is a digitally manipulated process that is based on layer coating of food by the appearance phase transitions or chemical reactions to solidify the layers. Culinary experience goes beyond taste, embracing all aspects, combining 3D printing techniques with help of which it is created personalized food with special design and layered structure. Currently 2D printers are specially designed to print with colorant jet food using cartridges that contain the basic colours and allow photographic quality reproduction of logos, images for customized decorations or photos. The print media may be various food sheets (wafer sheets, wafers, chocotransfer sheets) [1–4].

The 3D printing technique used in the food industry, in particular to obtain toppings for tortoises, candy that could not be created by other means, a major advantage of this technology by enabling it to control food allergies or food intolerance (gluten intolerance for example), helping to avoid the introduction of certain nutrients. Most 3D printers in the food industry work similarly to an ornate cornet, layers are gradually applied, and usually creamy ingredients such as chocolate, cheese cream, ice cream, mustard, peanut

G. I. Gheorghe (Ed.): ICOMECYME 2018, LNNS 48, pp. 56–61, 2019.
https://doi.org/10.1007/978-3-319-96358-7_6

cream are used. Also, 3D printing can not only be applied in the sweets area, it can print a mixture of wheat flour, so you can make custom pastes [3], (Fig. 1).

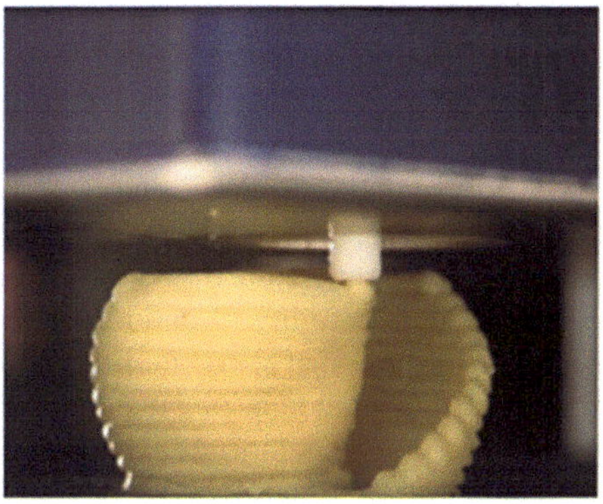

Fig. 1. Customized pasta realized by 3D printing [3]

An area of interest involves creating a type of food that can be used during long-term space missions. Consumption of powdered and oily foods but with a healthy nutritional intake can mark the end of the era in which people throw away the food because the 3D printing system uses a powder that has a shelf-life of decades and does not require keeping it at low temperatures. Therefore cartridges used in 3D printers containing sugars, complex carbohydrates and proteins or other essential organic substances can be used for a very long time. Since food powders can be used, calorific sources could be any commensurable thing that contains the right organic molecules. In the current growth rate of the Earth's population, the current food system will not be able to support 12 billion people, and once population numbers grow, we may not have the same agricultural resources [3].

Another aspect in which new 3D technology can help older people who have a healthy eating problem, with food being made from fresh food, especially vegetables, the food consumed after it has been recompressed by a 3D printer decomposes more easily into the cavity mouth than one prepared by classical methods. For example, peas, potatoes, chicken can be the basis of recipes for food preparation made by 3D food printers [3].

2 3D Printing Technology for the Food Industry

The platform on which the 3D printer is built in the article is based on the concept of 3D Hypercube printers, with a Cartesian XYZ axis system and a dosing system, all the process being controlled by a computer [2]. The composition of the food can be deposited/synthesized point by point or layer by layer according to the model created by the designer. The printhead may have various forms of construction, some use thermal energy (heating element, hot air) to sinter or melt the food powder, and others use inkjet spray heads that pulverizing binder or solvents. In this technology each layer is uniformly distributed over the manufacturing platform and a liquid binder is deposited on the print head to bind two consecutive layers of powder. The deposition material is stabilized by a fog of water and sugars and starch mixtures may be used. In the case of the depositing technology with the heating element, the depositing material is heated and then extruded through a depositing head and deposited on a metallic and cooled substrate, being especially used to create personalized chocolate or chocolate products. In the paper is presented a demonstrative model of the 3D printer with possibility of use in the food industry, offering more options for the dosing material so that the print head was designed especially from stainless steel provided with a system of resistors that can heat the chocolate raw material and a temperature transducer that allows rigorous control of the deposition material with the possibility of optimizing the computer-controlled manufacturing process, with the advantage of being compact and low maintenance costs. The disadvantage of this 3D printing method is long manufacturing time, visible deposition areas, delamination of layers due to temperature fluctuations (Fig. 2).

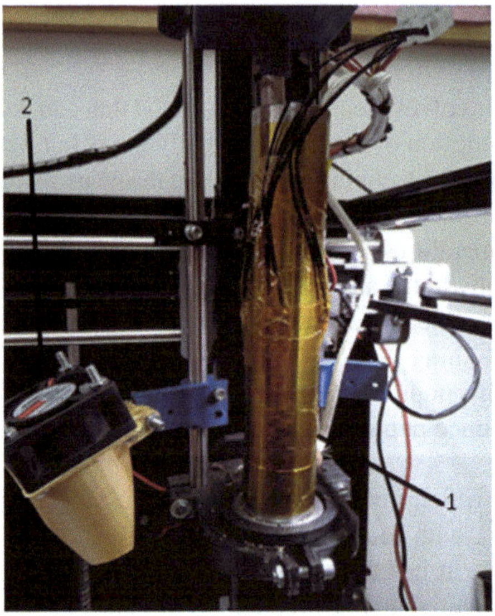

Fig. 2. Stainless steel head: 1 – deposition head with transducer; 2 – cooler

An important component of 3D printing technology using different printable materials is the deposition table that ensures the stability of the first layers and the basis of the future 3D food construction. In order to ensure an optimal adhesion temperature for various materials such as chocolate, sugar blends, potato puree (which can be classified as powdered food + binder), the deposition table was adapted by placing under the stainless steel plate of two PELTIER resistors which provide adequate cooling and ventilation capacity, Fig. 3.

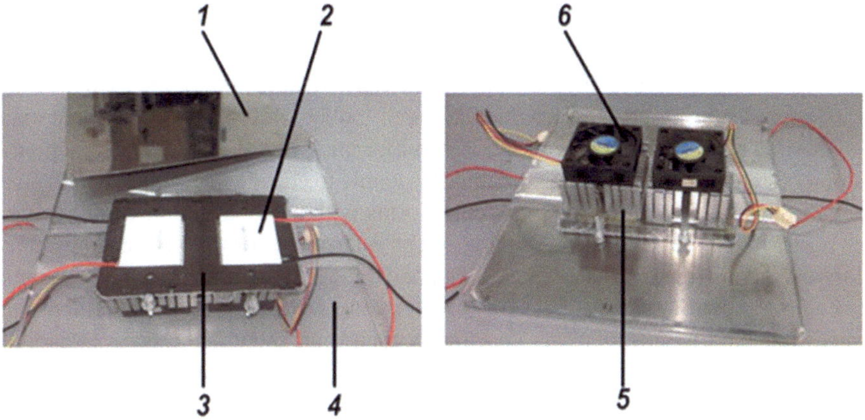

Fig. 3. Deposition table: 1 – stainless steel support plate; 2 – Peltier resistors; 3 – thermal insulator; 4 – support; 5 – aluminium radiator; 6 – ventilator

Some of the printable materials are stable enough to retain their shape after deposition, do not require further processing, and can be used in space applications. Other materials such as protein-based foods may require a post-deposition cooking process, this making it more difficult to keep the form. These materials used in traditional food recipes, additives and others are non-traditional edible materials (such as extracted from algae, sugar beet or even insects). The diversity of printing materials offers consumers the possibility of designing multi-material foods with complex geometries and structures [5, 6] (Fig. 4).

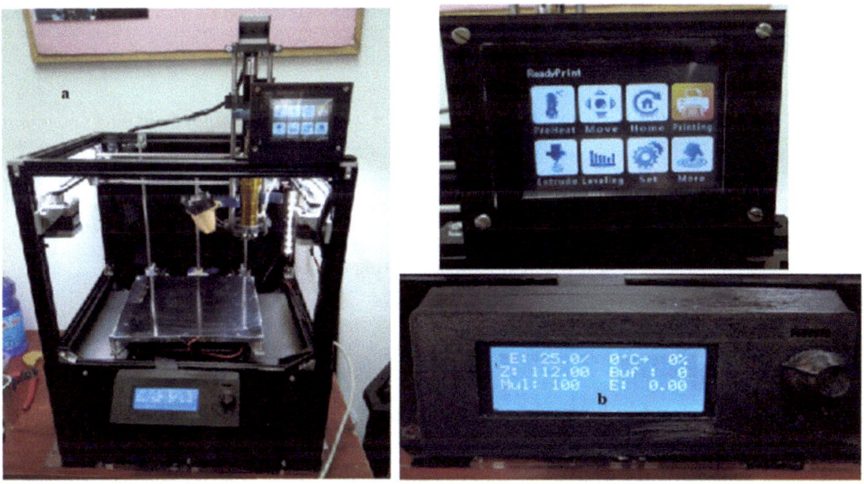

Fig. 4. 3D printer prototype used in the food industry: a – general view; b – electronic control panels

3 Impact of 3D Printing in the Food Industry

In addition to artistic appearance and custom food personalization for the food industry, the 3D manufacturing process provides research tools to develop new food materials. Also, this technology, which is in the beginning, must investigate the satisfaction of customers' requirements and the potential for people to change their lives. Foods made on 3D printers are realized in experimental forms with personalized flavors involving high costs. In the future, these technologies can offer more freedom in design, shapes, colors and flavors for home users. Foods obtained through this technology allow accurate date control, providing fresh and healthy preparations, well-known food ingredients and dosages, adapted to formulations specific to each manufacturing process. Also, 3D technologies allow us to reconsider our personalized food supply chain by bringing food to a shorter time for consumers. A barrier in the development of 3D food technologies is the composition of materials (ingredients and their structure, texture and taste) due to various combinations, handling under non-sterile conditions can damage the final product. Also an important condition is the feature of the printing materials that have to be rigid and strong enough to support the weight of the deposited layer. Key process parameters such as temperature, humidity, density, viscosity, thermal conductivity of foods should be taken into account in 3D printing of different types of food materials [1, 7] (Fig. 5).

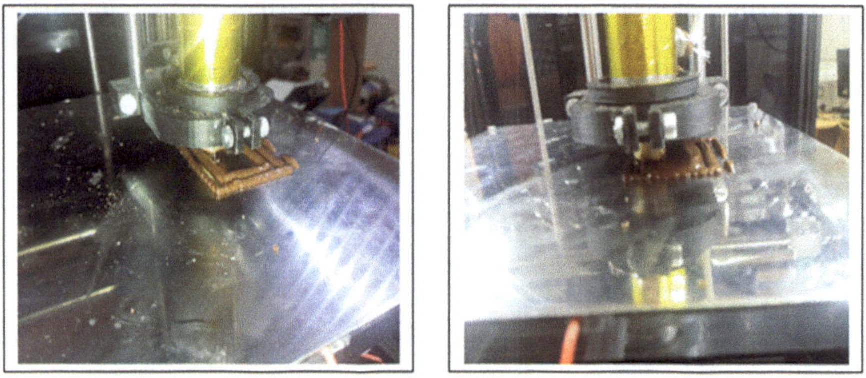

Fig. 5. Chocolate printing example

4 Conclusions

3D printing in the food industry has demonstrated the ability to print different homogeneous products used in the confectionery. However, this technology is initially with limited materials and structures, with the need to develop new printing materials, new design platforms and control rigorous nutrition. 3D printing in the food industry can be implemented in certain product development stages, new moulds and prototypes of sweets can be created, which often could not be created without a complex production line, thereby speeding up the product development process and reducing the cost of introducing a new product to the market. With the development of these technologies, this system can become environmentally friendly, with new ingredients, preparation of food on demand, collaboration between nutritionists for healthy diets. 3D Food Print Technology is a tool for researchers to use this technology in space for ease of transport and long storage of raw materials used. 3D technologies allow designers and users to use unique shapes and materials by offering high-quality products, fresh products, allowing users to develop new flavours, textures and shapes to create new culinary experiments [1].

Acknowledgements. This work has been funded by University POLITEHNICA of Bucharest, through the "Excellence Research Grants", Program UPB-GEX 2017. Identify: UPB-GEX2017, Grant no. 48/25.09.2017, ME 14-17-05, ID98.

References

1. Sun, J., Peng, Z., Zhou, W., Fuh, J.Y.N., Hong, G.S., Chiu, A.: A review on 3D printing for customized food fabrication. Procedia Manuf. **1**, 309–312, 316–317 (2015)
2. www.lesimprimantes3d.fr
3. http://3d4all.ro
4. http://www.novapan.ro
5. Izdebska, J., Żołek-Tryznowska, Z.: Agro FOOD Industry Hi Tech, vol. 27, no. 2, March/April 2016
6. Liu, Z., Zhang, M., Bhandari, B., Wang, Y.: 3D printing: Printing precision and application in food sector. Trends Food Sci. Technol. (2017). https://doi.org/10.1016/j.tifs.2017.08.018
7. Golding, M., et al.: Design & development of a 3-D food printer. Centre of Research Excellence, hosted by Massey University

Additive Technologies and Materials for Realization of Elastic Elements

Daniel Besnea(✉), Dana Rizescu, Ciprian Rizescu, Elena Dinu, Victor Constantin, and Edgar Moraru

Politehnica University of Bucharest, Splaiul Independenţei nr. 313, sector 6, Bucharest, Romania
d_bes@yahoo.com, {dana.rizescu,ciprian.rizescu}@upb.ro, elena_dinu@yahoo.com, victor.f.constantin@gmail.com, eddy_milan91@yahoo.com

Abstract. The paper presents a method of realization of representative types of elastic elements by means of additive technologies with the presentation of the main functional characteristics based on solid material deposition, FDM (Fused Deposition Modelling) technology, as well as the presentation of the main characteristics of the material used (PLA), highlighting domains and specific applications. Experimental results were obtained on a Hans Schmidt demonstration stand by comparing the deflections of various types of helical springs with circular, square and rectangular cross section.

Keywords: Additive technologies · 3D printing materials · Elastic elements

1 Introduction

Characterized by a wide spread across all compartments of mechanical engineering and mechatronics, in which many functional purposes are met, helical elastic elements have the advantages of occupying a reduced space in the assemblies in which they are integrated and achieving a relatively constant effect of force at a bigger stroke. Typically known as helical springs, they are made of wires or rods of different sections: circular, rectangular, square, trapezoidal, elliptical and annular, wrapped by a helix on a guide surface. Depending on the shape of the surface of the helix winding body, the helical springs may be cylindrical, conical, parabolic, hyperboloidal [1, 10].

The most common are cylindrical helical springs because of their manufacturing easiness. After the action force direction springs may be divided in: tension springs or compression springs. As regards the wire section this can be: a circular (most common), rectangular or square. The helical coiled spring load is mainly to the torsion but there is also a bending load, which may be disregarded if the wire angle α, it is small enough to $\alpha \approx (6° \text{ to } 9°)$. In the picture from Fig. 1 is presented a helical coiled spring subjected to tension load and spring with coils carried out.

G. I. Gheorghe (Ed.): ICOMECYME 2018, LNNS 48, pp. 62–70, 2019.
https://doi.org/10.1007/978-3-319-96358-7_7

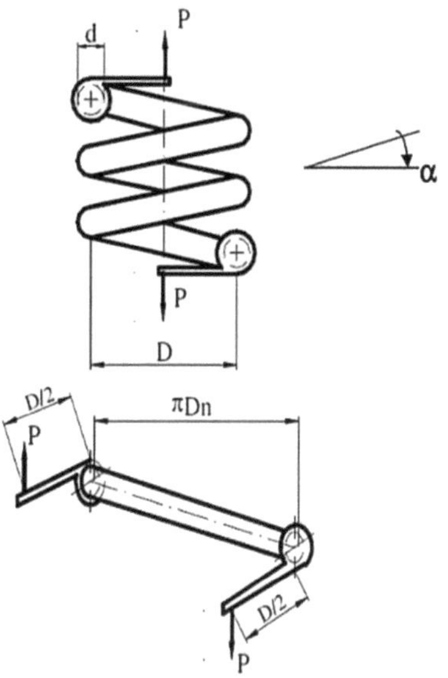

Fig. 1. Helical springs with tensile load

The acting force is labelled with P, the diameter of the spring wire with d, the number of wire turns is labelled with n and d is the medium diameter of the helical spring. In the case that angle α should be considered, there are two moments which load the helical coiled spring: a moment of torsion or torque - Mt, and a bending moment Mb.

$$M_t = \frac{P \cdot D}{2} \cdot \cos\alpha \tag{1}$$

$$M_b = \frac{P \cdot D}{2} \cdot \sin\alpha \tag{2}$$

One can observe that when $\alpha = 0$ then the torque $M_t = P \cdot D/2$ [1].

2 Materials and Fabrication Methods

The emergence of additive technologies in the early 1990s was a milestone in research and technology development. The new additive manufacturing (AM) technologies are the result of intense research and progress in various areas: from fine mechanics to numerical controls, from laser technology to three-dimensional modeling packs, from IT to material science. Rapid Prototyping technologies allow a great flexibility in application, an advantage to exploit micro components with a good dimensional precision used as conceptual model/functional prototypes or indirectly used as master models for the production of flexible tools for the manufacture of metallic or non-metallic parts in individual or small series production.

In recent years, a large number of innovative (Rapid Prototyping) technologies have been developed to transform the concept of achieving a complex product into a solid replica in a short period of time [2, 3]. Generally, these systems are a new class of virtual physical realization technologies using a family of special equipment. They provide the addition or bonding of material in successive sections as much as needed and where it is necessary. One of these additive technologies is FDM (Fused Deposition Modeling), a process based on the extrusion of material using a thread of different material qualities (polyamide, nylon, wax), which it heats up to a temperature a few degrees below the melting temperature, then reduces its diameter to 0.12–0.15 mm by extruding it into a depositing device, a device moving in the XOY plane to materialize a section of the 3D virtual model. The key of the process is to rigorous control the temperature at which the material is heated and maintained during the deposition [2].

The design step in the CATIA V5 software environment involves the execution of CAD models for the three types of helical springs with circular, square and rectangular cross section for all characteristic elements, namely helical arc height h = 50 mm, step p = 10 mm, Counterclockwise orientation, the Helix Curve Definition generation tool in the menu [4, 5] (Fig. 2).

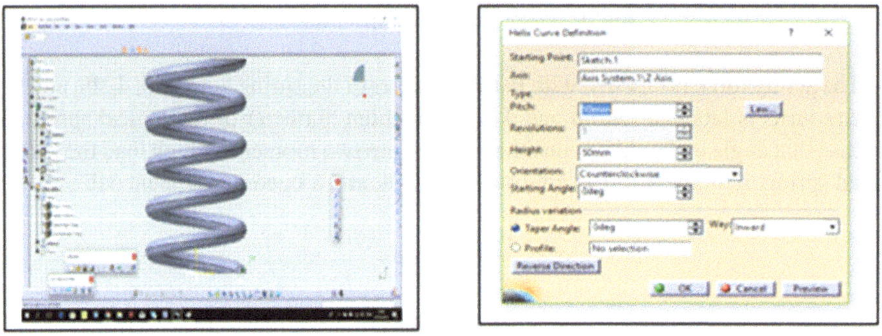

Fig. 2. Left – Generating of 3D model; Right – Generating of helical springs in CATIA with option Helix Curve Definition

A set of helical springs having a circular cross section of 4 and 8 mm diameter, rectangular section 4 × 2 mm and 4 × 8 mm and a square section with sides of 4 and 6 mm were realized (Fig. 3).

The manufacturing process using the FDM system comprises three main stages, namely the preprocessing stage, the proper construction stage of the part and the post-processing step. During the preprocessing stage the CAD model of the piece in .stl format is loaded, Fig. 4, designed in the CATIA V5 design environment in the QuickSlice specialized program - a program that generates the command code where the CAD model orientation in the work area of the depositing installation takes place so that the construction of the piece is optimal in terms of time work and material consumption.

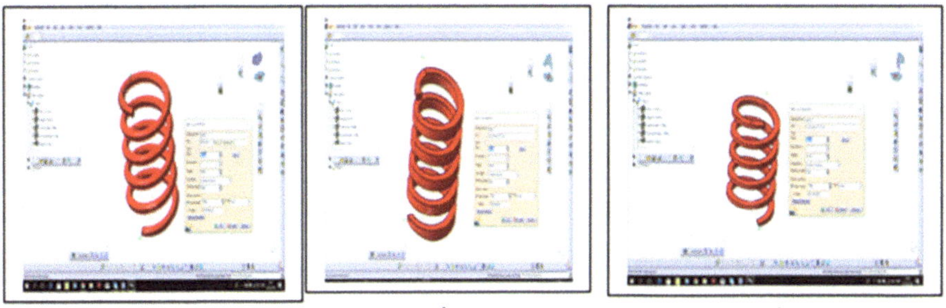

a b c

Fig. 3. Construction of the three types of helical springs: a – circular section; b – square section; c – rectangular section

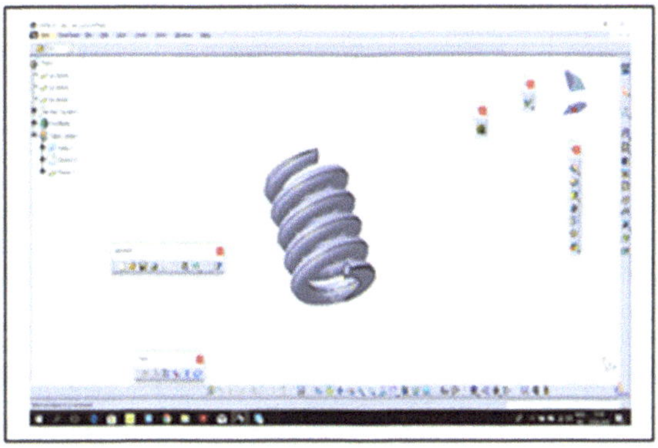

Fig. 4. The CAD model of the circular cross sectional helical spring saved in stl format

After the orientation of the CAD model, its sectioning is carried out with planes parallel to the plane of the machine (horizontal planes), operation resulting in several sets of level curves called perimeters. The sectional section along the Z axis is 0.2 mm is selected according to the diameter of the extrusion nozzle diameter, in the case presented in the article the diameter of the extrusion nozzles is 0.4 mm. The QuickSlice program [8] generates the paths the extruder needs to follow in order to materialize a section of the piece, Fig. 5 corresponding to the printer model and the Cura software used [7].

In the stage of construction of the piece, the layer with the layer is realized, the extrusion head of the machine deposits a thin thread of construction material along the curves defining the perimeter of the section and after the materialization of the perimeters, the deposition of the construction material takes place in the areas corresponds to the full areas of the piece, after the entire section is fully materialized, the platform descends with a step equal to the section of the virtual model and the entire process resumes for a new section until the last section of the virtual model of the piece is materialized, Fig. 6.

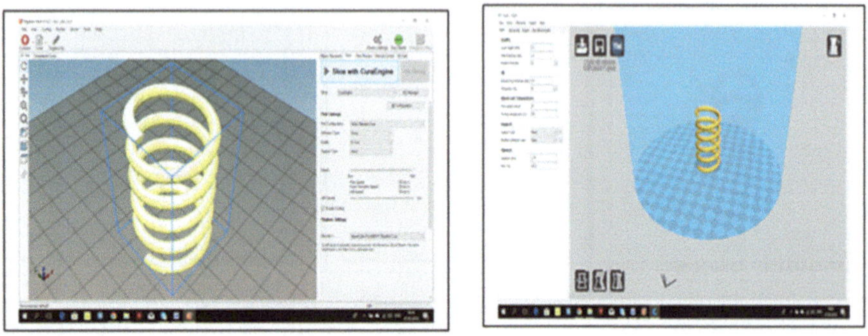

Fig. 5. Generate trajectories with QuickSlice and Cura software command interface of 3D Delta printing machine

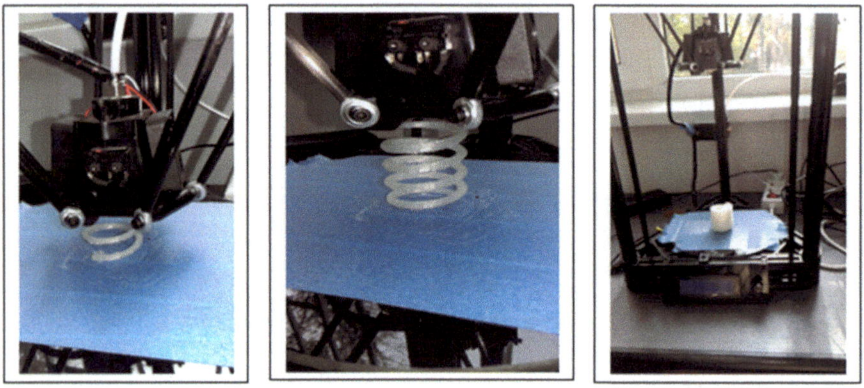

Fig. 6. Construction steps of different types of springs on the Delta printer

Delta FFF (Fused Filament Fabrication) printer has the following technical characteristics (Table 1) and due to the drive mechanism, the inertial forces are deviated 45° vertically and then discharged to pillars vertically. Thanks to this system and due to

Table 1. Technical characteristics of Delta printer

Naming	Value
Layer resolution	up to 50 μ
Max print speed	150 mm/s
Maximum print size	180 (diameter) × 300 (height) mm
Metal extruder	E3D V5 J-head
Positioning accuracy XYZ	0.01 mm
Nozzle diameter	0.4 mm
Filament	1.75 mm
Filament material	PLA (polylactic acid)

optimization of the software, the vibrations of the Delta printer are extremely small, the resonance is virtually non-existent and it can increase the speed by up to 4–5 times more than the classic Cartesian system printers [6].

Among the thermoplastic materials (solid plastics that become malleable by heating and hardened by cooling) used by FDM technology, was used as a base material the transparent PLA, from which some objects can be obtained at a slightly higher resolution weak compared to photopolymerizing resin, but with some superior structural properties. PLA is a bright, hard and biodegradable substance that chemically contains lactic acid and lactide. This has a low melting point - a temperature of 173–178 ℃ and tensile strength of 2.7–16 GPa with several application areas, including medical implants and packaging materials. PLA is mainly derived from corn starch and therefore ecologically. Resistant and more rigid than ABS, the PLA is more complicated to use in assembling parts that require bonding, and the deformation property at lower temperatures than the ABS (about 65 ℃) recommends it as the base material in making elastic elements (Fig. 7).

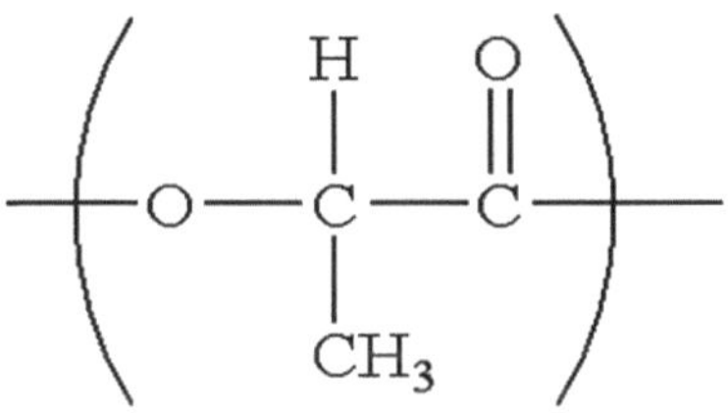

Fig. 7. PLA chain structure

Flexible material can be used for objects that are subject to stretches or compressive forces, from fashion design (e.g. a shoe, a frame of glasses) to engineering (a robot with multiple components that can withstand small shocks etc.). With 3D phosphorescent printing material, objects that light up in the dark can be created.

3 Experimental Results

In Fig. 8 is presented the experimental setup based on test stand, HV 500 N, which mainly consists of: 1 – distance measuring system, 2 – Imada force transducer, 3 – case for helical spring. Maximum testing force is 500 N. The Imada transducer is connected to PC for record the compression force and time [9] (Fig. 9).

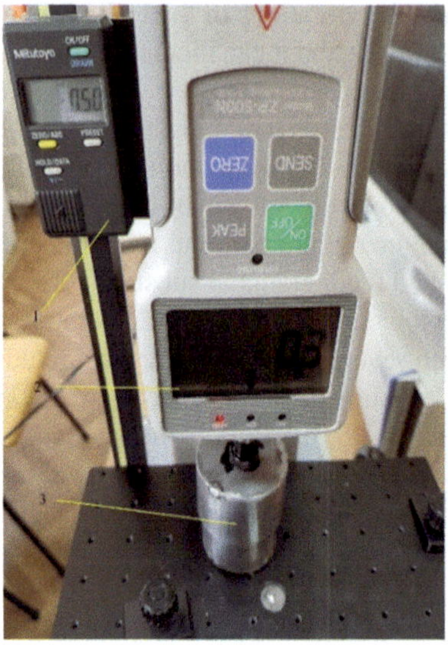

Fig. 8. Test stand Hans Schmidt [9]

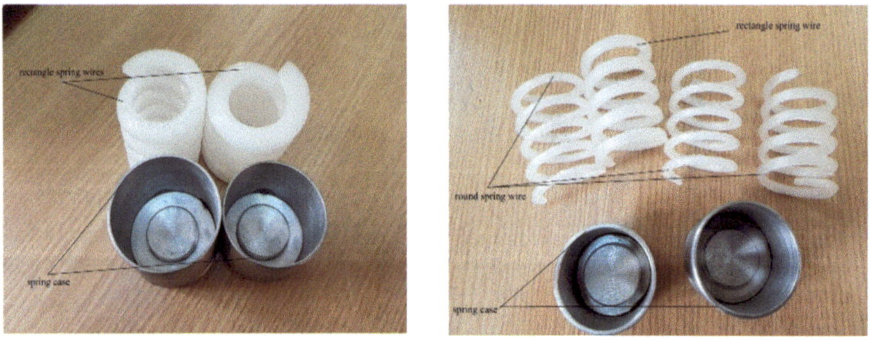

Fig. 9. Rectangle and circular spring wires realized by FDM (Fused Deposition Modeling)

In Fig. 10 are presented characteristics for square wire springs and in Fig. 11 for circular respectively. There were considered three helical springs with square wire PLA. For all these springs, there were determined the deflection and compression force as it is presented in Fig. 10. The wires has the dimensions: 3.5 mm, 5 mm and 7 mm. The preload deflections are print on each characteristic: $f01 = 6.96$ mm, $f04 = 2.17$ mm, $f05 = 0.382$ mm, which correspond to preload forces $P01 = 1.37$ N, $P04 = 2.47$ N, $P05 = 9.97$ N. The computed springs rigidities are: $c01 = 0.125$ N/mm, $c04 = 0.225$ N/mm, $c05 = 0.909$ N/mm.

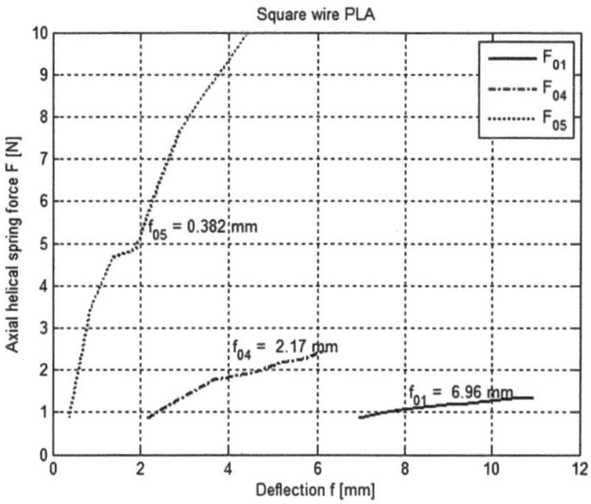

Fig. 10. The strokes of square wire springs

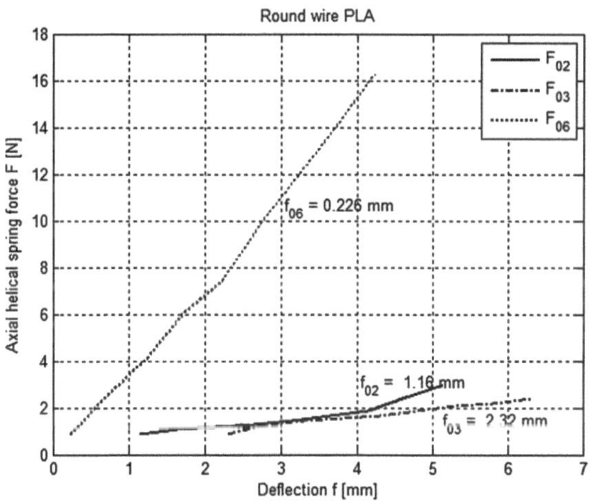

Fig. 11. The strokes of circular wire springs

Also, there were considered other three helical springs with circular wire PLA. For these springs, there were determined the deflection and compression force as it is presented in Fig. 11. The wires has the dimensions: 3 mm, 4.5 mm and 6 mm. The preload deflections are print on each characteristic: f02 = 1.16 mm, f03 = 2.32 mm, f06 = 0.226 mm, which correspond to preload forces P02 = 2.37 N, P03 = 3.87 N, P06 = 16.27 N. The springs rigidities are: c02 = 0.2162 N/mm, c03 = 0.271 N/mm, c06 = 1.484 N/mm.

4 Conclusions

Elastic elements made of plastics offer a number of advantages for spring applications. Lately, this area has grown strongly in response to increased demands of helical springs that combine the resistance of metallic materials with the special features of high-performance thermoplastics. However, it is important to understand when using plastic springs instead of traditional metallic springs. Plastic helical springs are designed to meet the specific requirements that involve the elastic element to be inert, non-corrosive and non-metallic. In this paper, helical springs with different sections of PLA thermo-plastic material have successfully been realised, the same 20% filling with help of FDM additive technology on a Delta printer. Also, the mechanical performances of the springs were studied by means of a Hans Schmidt test stand. Considering the characteristics square wire, the spring rigidities grow with wire dimension. The lowest rigidity corresponds to the spring labelled with 01. For round wire, the spring rigidities also grow with wire dimension. The lowest rigidity corresponds to the spring labelled with 03. The preload deflection is the biggest. All investigated characteristics are very close to a linear behaviour. The square wire springs have lower rigidities than round wire springs.

Springs made from PLA (polypolylactic acid), which is an ecological material with a high degree of biocompatibility; surely find its place in medical and pharmaceutical applications. In addition, with relatively good physical and mechanical properties, the PLA springs can arise as a reliable solution for aerospace, automotive, marine, instrumentation applications or where metal analogues can not be used for different reasons: water purification systems, chemical media, imaging equipment and radiography, etc.

Acknowledgements. This work has been funded by University POLITEHNICA of Bucharest, through the "Excellence Research Grants", Program UPB-GEX 2017. Identify: UPB-GEX 2017, Grant no. 48/25.09.2017, ME 14-17-05, ID98.

References

1. Demian, T., Plade, D.D., Curita, T.: Elemente elastice in constructia aparatelor de mecanica Fina. Editura Tehnica, Bucuresti (1994)
2. Berce, P., Balc, N., Caizar, C., Pacurar, R., Radu, A.S., Bratean, S., Fodorean, I.: Tehnologii de fabricatie prin adaugare de material si aplicatiile lor. Editura Academiei Romane, Bucuresti (2014)
3. Pandey, R.: Photopolymers in 3D printing applications. Degree Thesis Plastics Technology (2014)
4. Popovici, M.M.: Modelarea virtuala 3D in constructia de masini. Editura Printech, Bucuresti (2005)
5. Ghionea, I.G.: CATIA V5, Aplicatii in ingineria mecanica. Editura Bren, Bucuresti (2009)
6. www.build3dparts.com
7. https://ultimaker.com
8. Repetier-host V1.6.2 - QuickSlice software
9. https://www.hans-schmidt.com/en/produkt-details/test-stand-hv-500n/
10. Rizescu, C.I.: Elemente si mecanisme de mecanică fină - Partea I. Editura Printech, Bucuresti (2013)

Analytical and Experimental Studies on Wear Behaviour of Cast and Heat Treated $AlSi_{12}CuMgNi$ and $AlZn_6MgCu$ Matrix Composites Reinforced with Ceramic Particles, Under Sliding Conditions

Ileana Nicoleta Popescu, Ivona Camelia Petre(✉), and Veronica Despa

Valahia University of Targoviste, 13 Aleea Sinaia Street, Targoviste, Romania
petreivonacamelia@yahoo.com

Abstract. The working conditions of the composite materials used to produce machine parts lead to different forms of wear. The fact that, for example, for a kinematic coupling with sliding motion is often used a material with higher hardness (cast iron, steel) in combination with a material with a lower hardness (a composite material) there is the possibility of wear through abrasion and local plastic deformations. The paper proposes an analytical model for the determination of wear, depending on the angle of inclination of the roughness of the hard surface. The experimental wear investigations were made on cast iron disc (300 HB hardness) at room temperature using a "pin on disc" machine, at $3.5 \cdot 10^{-1}$ MPa and 7.5×10^{-1} MPa contact pressure and 3.8 m/s sliding speed. The composite consisted from cast and heat treated $AlSi_{12}CuMgNi$ and $AlZn_6MgCu$ matrix reinforced with Al_2O_3 and Graphite combined in different proportion, in the 0–5 volume percent range. The experimental results of the wear for the different materials are analyzed and compared to the analytical ones. The comparison of the experimental and the theoretical results confirms the veracity of the model and corresponds with many of the experimental results obtained in the specialized works.

Keywords: Wear behaviour · Aluminium matrix composites
Ceramic reinforcement

1 Introduction

The wear process, consisted in material separation followed by the change in the initial condition of the friction surfaces, results in wear of the friction coupler. Due to the degradation of the surfaces in contact there is a change in the dimensions of the respective parts as well as in the operation of the friction coupler, which is not desirable. From the tribological point of view, an optimal correlation should be found in the selection of the friction coupler materials, the surface treatment mode and the loading and exploitation conditions. Metallic matrix composites (Al, Cu, Fe, etc. matrix) reinforced with soft ceramic particles (Graphite, MoS_2 etc.) have the objective to: (i) stabilize the developed friction coefficient during braking, particularly at high temperatures; (ii) decrease wear

G. I. Gheorghe (Ed.): ICOMECYME 2018, LNNS 48, pp. 71–82, 2019.
https://doi.org/10.1007/978-3-319-96358-7_8

of disc and also (iii) increase grippe resistance [1–4]; (iv) exhibit a low friction coefficient (compared to composites reinforced with hard ceramic particles [5–8]) are mostly used in tribological applications without lubrication. We could say that this type of material is a balance of mechanical resistance with reduced friction and wear. Thanks to these benefits, these materials are often used in the construction of various kinematic couplers with sliding motion, because besides the benefits presented above they have a specific low density and environmental resistance. The tribological characteristics of the materials can be improved by adding both hard and ceramic reinforcement components (SiC, TiC, Al_2O_3, SiO_2, ZrO_2, or Mullite) in combination with soft ones (Graphite, MoS_2) [9–12], which recommends them as new solutions for the vehicle braking system. Used in combination with gray cast iron as a sliding-engagement couple, they have good thermal conductivity (required for proper heat removal due to friction) [13–16]. By analyzing the tribological behavior of couplings "pin" (Al composites) on "disc" (cast iron), we follow the evolution of the wear intensity according to pressure, speed, mode of surface processing).

This paper proposes a theoretical model of wear characterization and comparison of the obtained results with the experimental ones.

2 Analytical Model for Calculation of Wear Composite Material

As it is known from the literature [10, 11, 17–29] the appearance of a wear or other shape depends, in particular, on the ratio of the hardness of the surfaces in contact, the surface processing mode, the contact pressure between the conjugated surfaces, the exploitation conditions, etc. Abrasion wear is a form of wear present in most slip-resistant couplings with essentially different hardness but is also accompanied by other types of wear depending on the kinematic coupling specificity. Wear generally has the effect of reducing the size of the parts and is accompanied by the separation of the wear particles. The experimental researches and the theoretical investigations carried out by Khruschov [26, 27], Koplalinsky [28], Huq [29] have shown that the basic law of abrasive particles detachment is of the form:

$$I_{uv} \approx \frac{U_v}{L_f} \tag{1}$$

here: I_{uv} is the volumetric wear intensity; U_v is the volume of waste material; L_f is the length of friction. If the contact surface of the contacting kinematic coupler remains constant over the wear period (A_n nominal area) and the thickness of the worn layer is uniform then:

$$I_{uv} = \frac{U_v}{L_f} = \frac{A_n \cdot d}{L_f} = A_n \cdot I_{uh} \tag{2}$$

Where: d is the depth of wear; I_{uh} is the linear wear intensity. If the friction surfaces of the coupling elements have different hardnesses, micro-scraping processes may occur in the contact area. Under the same friction path and the same nominal contact pressures, the thickness of the coat depends on the surfaces hardness. To formulate this hypothesis, many researchers [29–32], consider the Archard wear law, according to which the wear volume U_v is directly proportional to the normal force (F_n) and the friction length (L_f) and inversely proportional to surface hardness (H):

$$U_v = K\frac{F_n \cdot L_f}{H} \quad (3)$$

Where: K is the proportionality factor. The proportionality factor is considered as a wear ratio (Archard) and a material characteristic.

The values of this coefficient are always subunits, usually $10^{-3} \ldots 10^{-9}$ of the order of magnitude.

3 Experimental Part

3.1 Selection of Materials

Discontinuous particle reinforced Metal Matrix Composites (MMCs) are particularly attractive because in addition to their low cost of manufacture, they can be shaped by conventional metalworking processes [6–13, 33]. This has led to increased interest in the potential for large-scale use of particle reinforced MMCs in the automotive industry as materials for pistons, connecting rods, brake rotors, calipers and liners [34–36].

The wear behavior of Al_2O_3 particle reinforced MMCs has been extensively studied by various workers [5–8] and the tribological characteristics of Al-Si/Graphite for cylinder liners was studied in detail by Krishnan et al. [34]. The wear resistance of particulate reinforced MMCs depends mainly on the particle size, particle volume fraction and matrix alloy properties [1–13]. Al–Si alloys are commonly used in the manufacture of automotive engine components such as cylinder blocks, pistons and piston insert rings. The principal reasons for the usage of Al–Si alloys are their good castability, high corrosion resistance [35] and low density. The wear resistance of Al–Si alloys is strongly dependent on the alloy composition, applied load and sliding speed [4, 34]. The wear resistance of these alloys can be enhanced by the incorporation of a hard ceramic phase in the soft Al alloy matrix. On base of good physical and mechanical characteristics of hardening by thermal treatment of $AlZn_6MgCu$ alloys [37, 38] we choose these alloys as matrix. Graphite (Gr), in the form of fibers or particulates, has long been recognized as a high-strength, low-density material [1–4]. Al/Gr particulate MMCs produced by solidification techniques represent a class of inexpensive tailor-made materials for a variety of engineering applications such as automotive components, bushes, and bearings. Their uses are being explored in view of their superior technological properties such as the low coefficient of friction, low wear rate, high damping capacity and good machinability [34–36].

Al_2O_3 particles are the most commonly used reinforcements in MMCs and the addition of these reinforcements to aluminium alloys has been the subject of a considerable amount of research works [7]. Preparation and casting of material composites was presented in details in a previous work [9]. The heats treated of composites [9] consisted in: (i) Solution treatment of 480 ± 5 ℃/60 min and quenched in water and artificial ageing at 140 ± 5 ℃/12 h, air quenched, for $AlZn_6MgCu/(Gr, Al_2O_3)$ materials; (ii) Solution treatment of 520 ± 5 ℃/6 h and quenced in water and artificial ageing at 170 ± 5 ℃/14 h, air quenched, for $AlSi_{12}CuMgNi/(Gr, Al_2O_3)$ materials.

3.2 Mechanical Testing

The mechanical test was Brinell hardness according with Romanian standards STAS201/2 for $AlSi_{12}CuMgNi$ matrix composites and STAS7608/88 for $AlZn_6MgCu$.

3.3 Tribological Experiments

We determined the average wear rate for all types of composites and unreinforced matrix in cast state. The disc was polished before each tribological test. The tribological tests were performed on the pin-on-disc type wear machine, under following condition: contact pressure (applied load) $3.5 \cdot 10^{-1}$ MPa and 7.5×10^{-1} MPa, relative sliding speed of 3.8 m/s at constant temperature, only for heat treated materials.

All tests were run under dry sliding conditions. The dimensions of all samples was 10 × 10 × 10 mm. The height (mm) of the worn layer was measured after sliding the pin-on-disc. The obtained results were the average of three–four measurements and the time of wear test was kept constant in all cases at 30 min.

3.4 Mechanical Characteristics

The Brinell hardness was measured for the two types of composites (Al-Si based, and Al-Zn based materials). Figure 1(a) shows the hardness of $AlSi_sCuMgNi$ matrix composite reinforced with particles of graphite and Al_2O_3, in varying proportions, for both untreated and treated composite materials. Figure 1(b) shows the hardness of the $AlZn_6MgCu$ matrix composite under the same experimental conditions.

It is obvious that for composite materials with a higher graphite percentage the hardness decreases. After heat treatment of aluminium based composites, age hardening appears leading to increasing of hardness. These aspects are confirmed by experimental results when for the materials to which treatments have been applied the hardness is higher than the untreated ones. In the case of materials reinforced with Al_2O_3, we observe that the higher the percentage of Al_2O_3, leading to the higher of hardness. Our results confirmed that the materials with hard ceramic particles (Al_2O_3) in composition are harder than those with soft ceramic (Gr) particles in composition.

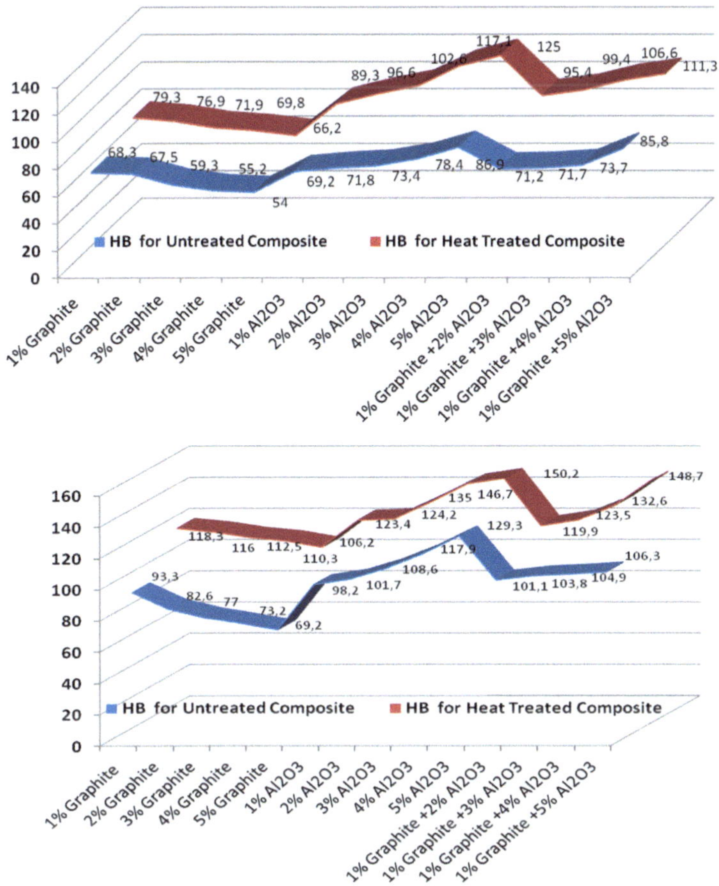

(a)

(b)

Fig. 1. The average values of Brinell Hardness (HB) for (a) $AlSi_{12}CuMgNi$ and (b) for $AlZn_6MgCu$ matrix composites in untreated and heat treated conditions.

3.5 Tribological Results

Wear (loss in height/the depth of the lost layer [mm/h]) ns for heat treated $AlSi_{12}CuMgNi$ (see Fig. 2) and $AlZn_6MgCu$ (see Fig. 3) respectively reinforced with graphite or/and Al_2O_3 particles for 0.35 MPa and 0.75 MPa (see Figs. 2 and 3).

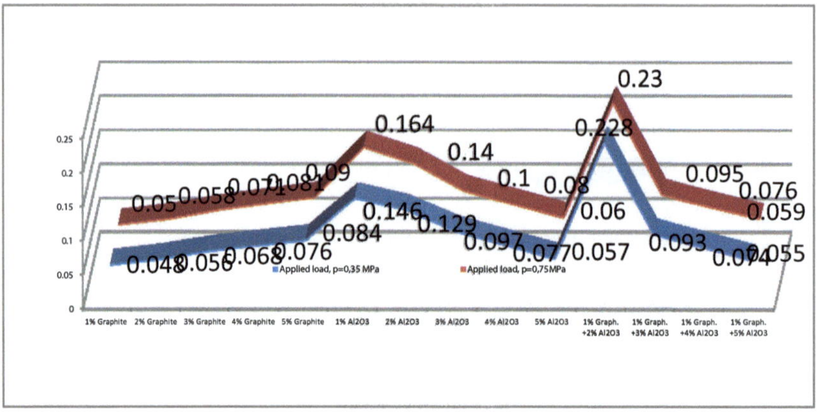

Fig. 2. The average values of determined wear [mm/h]) for $AlSi_{12}CuMgNi$ based composite depending on different proportion (%vol.) of the Graphite or/and Al_2O_3 reinforcement particles.

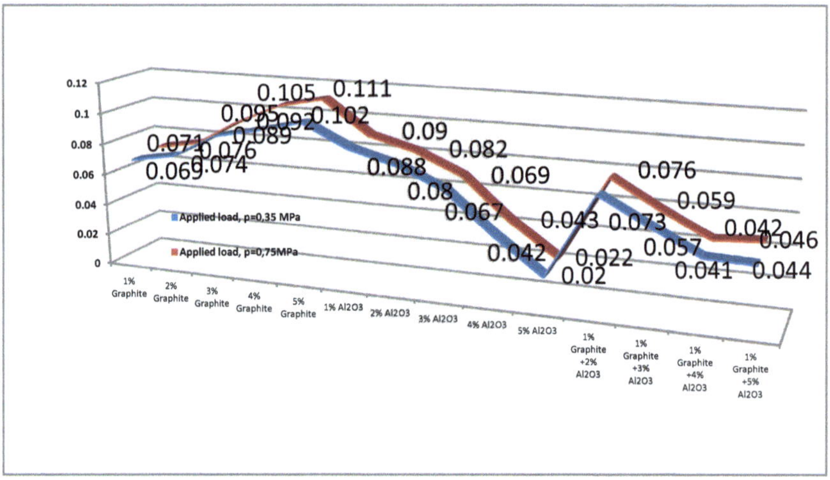

Fig. 3. The average values of determined wear [mm/h]) for $AlZn_6MgCu$ based composite depending on different proportion (%vol.) of the Graphite or/and Al_2O_3 reinforcement particles

3.6 Comparison of Calculated Volume of Wear, Based on the Theoretical Model, with the Results Obtained Experimentally

For some known working conditions (hardness, loading, friction length, etc.) the theoretical volume of wear can be determined. Figure 4 shows the theoretical evolution of the volume of waste material (for $AlZn_6MgCu$/Gr) for a friction length $L_f = 7 \cdot 10^7$ mm, a load of $F_n = 0.35$ MPa, hardness $H = 70 \ldots 95$ HB and various proportional value (K) values.

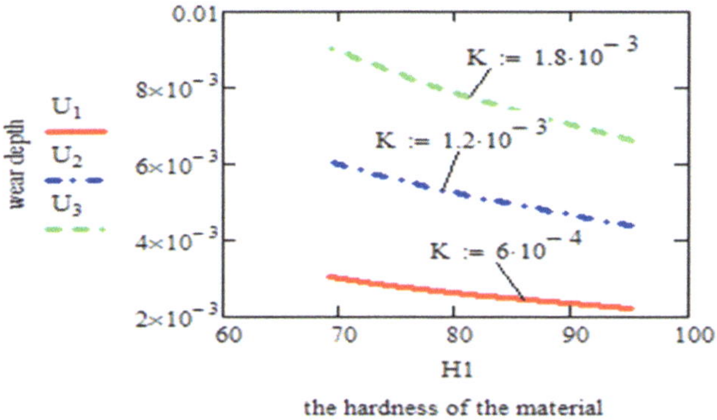

Fig. 4. Theoretical evolution of used material volume for $AlZn_6MgCu/Gr$ for known conditions

It is worth noting that the theoretical volume of wear decreases with increasing the material hardness. If replacements are made then the volumetric wear intensity:

$$I_{uv} = K\frac{F_n}{H} \quad (4)$$

As mentioned, the processing of the surfaces in contact has a decisive influence on the evolution of the wear intensity. It is considered a rigid surface (1) at which the angle of attack of the cone (inclination of the asperities) is α, which moves at a constant speed on a deformable plastic plane (2), characterized by the shear resistance τ_f.

In this case, the ratio between the volume of deformed and removed material (U_v) and the friction length (L_f) under normal load (F_n) [28, 31, 39–41] is:

$$K_A = \frac{U_v}{L_f \cdot F_n} = \frac{1}{2 \cdot \tau_f}\frac{(\sin\alpha)^2 + 0.5 \cdot \sin 2\alpha}{1 + \sin\alpha} \quad (5)$$

Accepting Tabor's hypothesis that the shear resistance (τ_f) of a material is dependent on the hardness of the material (H):

$$\tau_f = \frac{H}{3\sqrt{3}} \quad (6)$$

the Archard wear coefficient (K) can be determined. For these reasons, the wear volumetric intensity will be:

$$I_{uv} = \frac{2 \cdot tg\alpha}{\pi \cdot H} \cdot F_n \quad (7)$$

Figure 5 shows the evolution of the wear intensity for one of the analyzed materials ($AlZn_6MgCu$ reinforced with Al_2O_3), the loading function ($p = 0.35$ MPa) and the angle of inclination of the rigid cone. From the graph analysis in Fig. 5 it is worth noting that for a certain inclination of the conical penetrator the wear intensity increases slightly, and then there is a rapid increase of the wear. This indicates that for rigid cone angles lower than 55°–60° a moderate wear is occurring, after that, for larger angles 60°–65°, a severe wear occurs. The explanation of this is due to the response behavior of the material with lower hardness under the action of the conical penetrator. At small angles, the material deforms in the form of a wave (without detachment of particles) by micro-ploughing, and at higher angles then the 60° micro-cracking of the microcutting material takes place (see Fig. 6). The transition from waveform deflection to micro-deflection occurs when the tilt angle of the rigid cone exceeds a critical value (α_{cr}), the value which depends on the characteristics of the studied material [18]. This theoretical evolution (depending on the processing of the surface of the materials) corresponds to many of the experimental results obtained in the specialized works [18, 42–44]. Figure 7 shows the evolution of wear intensity calculated for $AlZn_6MgCu$ with 2% Gr and 2% Al_2O_3 respectively, for $p = 0.75$ MPa).

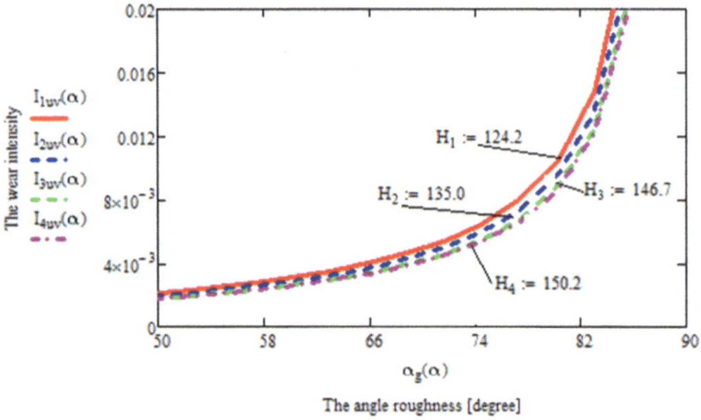

Fig. 5. Evolution of wear intensity with hardness of $AlZn_6MgCu/Al_2O_3$ composite (Ii.uv(α), depending of loading and tilt angle of rigid cone.

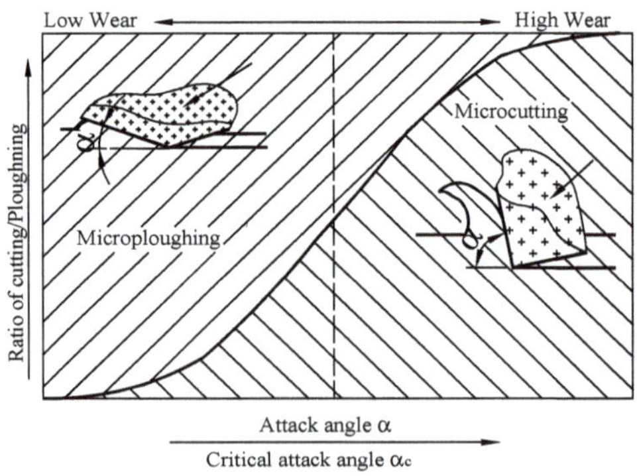

Fig. 6. Ratio of microcutting to microploughing as a function of the ratio of the attack angle to the critical attack angle.

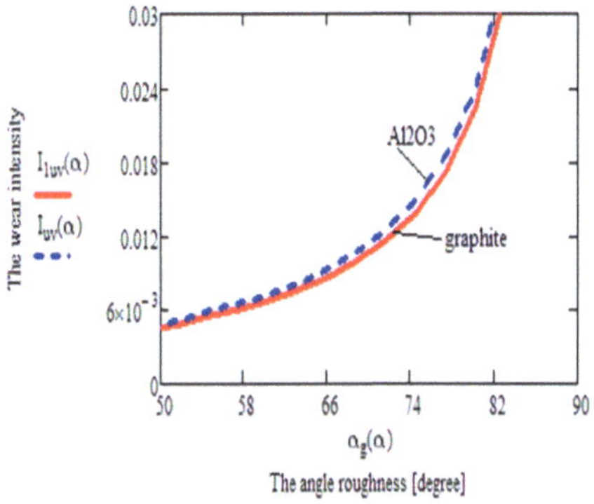

Fig. 7. Evolution of wear intensity depending on the angle of attack cone and reinforced material

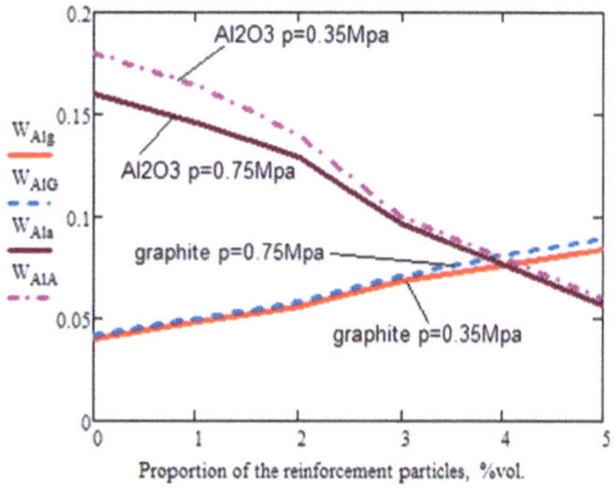

Fig. 8. Evolution of wear layer with pressure and reinforcement (%vol), for $AlSi_{12}CuMgNi$ with graphite or Al_2O_3

Figure 8 shows the evolution of wear depth for $AlSi_{12}CuMgNi$ reinforced with graphite or Al_2O_3 particles for the two working pressures 0.35 MPa and 0.75 MPa. It is noteworthy that for $AlSi_{12}CuMgNi$ the depth of the wear layer increases with increasing graphite content and decreases as the content of Al_2O_3 increases. Figure 9 present the theoretical and experimental results of the volume of waste material, obtained for $AlZn_6MgCu$ reinforced with graphite particles (Fig. 9a) and Al_2O_3 (Fig. 9b). It is noted

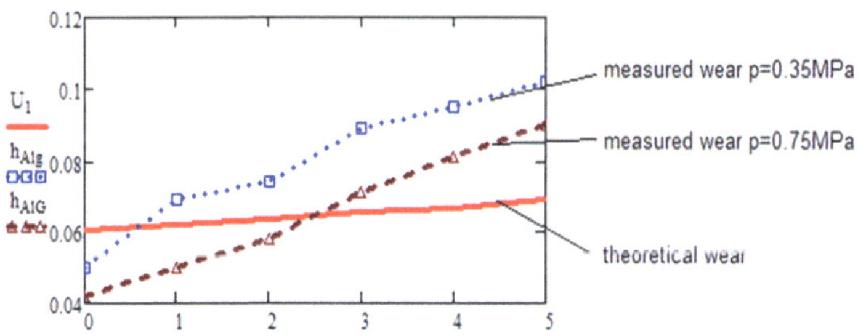

(a)

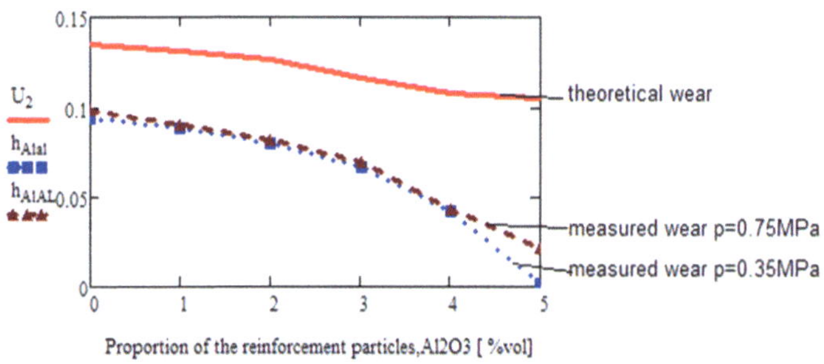

(b)

Fig. 9. Evolution of volume of material used with pressure and proportion of (a) Graphite (% vol), respectively (b) Al_2O_3 (%vol), for $AlZn_6MgCu$ based composite.

that for low values of reinforced materials percentage (0%–2%), the results are not essentially different.

4 Conclusions

Knowing the height (depth) of the used material, the dimensions of the samples and the length of the friction, under experimental conditions, it is possible to determine the volume of used material that will be compared to that obtained theoretically. In the present paper we also found the following: (a) under the experimental conditions it was not possible to highlight the way of processing the material due to the small dimensions of samples. In this case, we can't compare the obtained values; (b) the theoretical model has highlighted the fact that the way of processing the friction surfaces (materialized by the angle of inclination of the asperities) influences the wear. For tilting angle values a mild wear occurs, above this value wear significantly increases; (c) the increase of the wear intensity according to the angle of machining of the asperities is explained by the fact that at the angles there is the deformation of the material without detachment of microploughing particles, and at higher angles there occurs microforming deformation of the material, which makes that wear will grow rapidly; (d) in the theoretical model it was emphasized that with the increase of the graphite content the wear increases, which is also confirmed by the experiments; (e) comparing the theoretical and experimental results, it has been found that samples made of embedded graphite materials wear out more than those with the same processing and loading conditions (pressure, speed); (f) comparing the theoretical model with the experimental results it was confirmed that wear decreases with increasing material hardness; (g) in the experiments we can see the increase of the depth of the used layer with the load; (h) under experimental conditions also, for $AlSi_{12}CuMgNi$ and $AlZn_6MgCu$ analyzes, the depth of the wear layer increases with increasing graphite content and decreases as the content of Al_2O_3 increases.

References

1. Anderson, A.E.: Friction and wear of automotives brakes. In: ASM Handbook: Powder Metallurgy, vol. 7, pp. 569–577.ASM International, Materials Park (1984)
2. Chan, D., Stachowiak, G.W.: Review of automotive brake friction materials. Proc. Inst. Mech. Eng. Part D: J. Automob. Eng. **218**, 953–966 (2004)
3. Popescu, I.N., Ghiţă, C., Bratu V., Palacios Navarro, G.: Tribological behaviour and statistical experimental design of sintered iron-copper based composites. Appl. Surf. Sci. **285**(Part A), 72–85 (2013)
4. Furlan, K.P., da Costa Gonçalves, P., Consoni, D.R., et al.: Metallurgical aspects of self-lubricating composites containing graphite and MoS_2. J. Mater. Eng. Perform. **26**(3), 1135–1145 (2017)
5. Petre, I.C., Popescu, I.N.: The phenomenological analysis of the nature of the friction, from theoretical and experimental point of view of Al-Al_2O_3-graphite composite/cast iron "pin on disc" sliding system. Int. J. Mechatron. Appl. Mech. (2), 40–47 (2017)

6. Popescu, I.N., Vidu, R., Bratu, V., Olei, A.B., Ungureanu, D.N., Anghelina, F.V.: Effects of silicon carbide proportion and artificial aging parameters on microstructure and hardness of Al-Cu/SiCp composites. Solid State Phenom. **216**, 122–127 (2015)
7. Yang, L.J.: The transient and steady wear coefficients of A6061 aluminium alloy reinforced with alumina particles. Compos. Sci. Technol. **63**(3–4), 575–583 (2003)
8. Popescu, I.N., Zamfir, S., Bojin, D., et al.: Physical and quantitative microstructural analysis of sintered Al-Cu/SiCp composites. Mater. Sci. Forum **672**, 251–254 (2011)
9. Popescu, I.N., Bratu, V., Filip, V., Catangiu, A., Ungureanu, D.N., Anghelina, F.V.: Tribological and mechanical behavior of cast Al particulate composites and hybrid Al matrix composites. Sci. Bull. Valahia Univ. Mater. Mech. **4**, 109–116 (2009)
10. Kaushik, NCh., Rao, R.N.: Influence of applied load on abrasive wear depth of hybrid Gr/SiC/Al–Mg–Si composites in a two-body condition. J. Tribol. **139**(6), 061601 (2017)
11. Sima, G., Mangra, M., Gîngu, O., Criveanu, M.C., Olei, B.A.: Influence of the reinforcing elements on the wear behavior of Al/(SiC+graphite) composites elaborated by spark plasma sintering technology. Mater. Sci. Forum **672**, 241–244 (2011)
12. Popescu, I.N., Zamfir, S., Anghelina, F.V., Rusanescu, C.O.: Fabrication through P/M of ecological aluminum based composite materials. Part 1-Characterization and densification of mixture powders. In: Proceedings of the 2nd International Conference on MEQAPS Constantza, pp. 200–205 (2010)
13. Tandon, K.N., Feng, Z.C.: Wear behaviour of SiC part reinforced Al composites against steel balls under dry and lubricated condition. Tribol. Lett. **6**, 113–122 (1999)
14. Petre, I., Stoian, E.V., Enescu, M.C.: Determining the heat regime in the working of a coupling with sliding motion. Sci. Bull. Valahia Univ. Mater. Mech. **14**(11), 33–38 (2016)
15. Rusănescu, C.O., Rusănescu, M., Anghelina, F.V.: Variation of mechanical properties with temperature for an ecomaterial. Optoelectron. Adv. Mater. – Rapid Commun. **7**(11–12), 947–951 (2013)
16. Tudor, A.: Frecarea şi uzarea materialelor, pp. 158–179. Editura Bren, Bucuresti (2002)
17. Tudor, A., Vlase, M.: Uzarea materialelor, pp. 232–246, Editura Bren, Bucureşti (2010)
18. Coronado, J.J.: Effect of abrasive size on wear. In: Marcin, A. (ed.) Abrasion Resistance of Materials, pp. 167–184. InTech, Janeza Trdine (2012)
19. Ghita, C., Pop, N., Popescu, I.N.: Existence result of an effective stress for an isotropic visco-plastic composite. Comput. Mater. Sci. **64**, 52–56 (2012)
20. Olei, B.A., Ştefan, I., Popescu, N.: The influence of the sintering temperature on the wear testing for some steels samples obtained by powder metallurgy. Solid State Phenom. **216**, 216–221 (2014)
21. Petre, I., Popa, C., Dumitru, D., Manescu, C.: Analytical model of calculus for influence the translation guide wear over the machining accuracy on the machine tool. Fiability & Durability/Fiabilitate si Durabilitate **2**(6), 17–22 (2010)
22. Petre, I., Poinescu, A.A., Catangiu, A., Mihai, S.: Studies regarding the reaction method to wear braking mechanism. Sci. Bull. Valahia Univ. Mater. Mech. **15**, 37–42 (2017)
23. Bratu, V., Popescu, I.N., Stoian, E.V., Ungureanu, D.N., Rusănescu, C.O., Toma, L.G., Voicu, A.C.: Casting and homogenization of $AlCu_3.3Mg_{1.5}Mn$ Al alloys for aircraft industry. Adv. Mater. Res. **1128**, 44–50 (2015)
24. Popescu, I.N., Zamfir, S., Anghelina, V.F., Rusănescu, C.O.: Processing by P/M route and characterization of new ecological Aluminum Matrix Composites (AMC). Int. J. Mech. **4**(3), 43–52 (2010)
25. Rusănescu, C.O., Rusănescu, M., Ioedănescu, T., Anghelina, F.V.: Mathematical relation ships between alloying elements and technological deformability indexes. J. Optoelectron. Adv. Mater. **15**(7–8), 718–723 (2013)

26. Khruschov, M.M.: Resistence of Metals to Wear by Abrasion as Related to Hardness. Institution of Mechanical Engineers, London (1957)
27. Khruschov, M.M.: Principles of abrasive wear. Wear **28**, 69–88 (1974)
28. Kopalinsky, E.M., Oxley, P.L.: Explaining the mechanics of metallic sliding friction and wear in terms of slipline field models of asperity deformation. Wear **190**(2), 145–154 (1995)
29. Huq, M.Z., Celis, J.-P.: Expressing wear rate in sliding contacts based on dissipated energy. Wear **252**, 375–383 (2002)
30. Kato, K.: Classification of wear mechanisms/models. J. Eng. Tribol. **216**(6), 349–355 (2002)
31. Hutchings, I., Shipway, P.: Tribology: Friction and Wear of Engineering Materials, pp. 174–217. Elsevier, New York (2017)
32. Zmitrowicz, A.: Wear patterns and laws of wear. J. Theor. Appl. Mech. **44**, 219–253 (2006)
33. Popescu, I.N., Bratu, V., Ionescu, M., Chivu, M., Enescu, M.C., Poinescu, A.A.: Preparation and characterization of cast aluminium/graphite composites and hybrid aluminium composites. Sci. Bull. Valahia Univ. **4**, 104–108 (2009)
34. Krishnan, B.P., Raman, N., Narayanaswamy, K., et.al.: Performance of an Al-Si graphite particle composite piston in a diesel engineering. Wear **60**, 205–215 (1980)
35. Popescu, I.N., Enescu, M.C., Bratu, V., Zamfir, R.I., Stoian, E.V.: Development, microstructure and corrosion resistance of Al-Mg-(Si) binary and ternary system samples in 5.3% NaCl solution for applications with environmental impact. In: Advanced Materials Research, vol. 1114, pp. 239–244. Trans Tech Publications, Switzerland (2015)
36. Prasad, S.V., Asthana, R.: Aluminum metal matrix composites for automotive applications: tribological considerations. Tribol. Lett. **17**(3), 445–453 (2004)
37. Enescu, M.C., Popescu, I.N., Zamfır, R., Molagıc, A., Bratu, V.: Influence of heat treatment on microstructure and corrosion behavior of 7xxx Al alloys. In: Proceedings of the 2nd International Conference on MEQAPS, pp. 212–216 (2010)
38. Ibrahim, I.A., Mohamed, F.A., Lavernia, E.J.: Particulate reinforced metal matrix composites, a review. J. Mater. Sci. **26**(5), 1137–1156 (1991)
39. Challen, J.M., Oxley, P.L.B.: An explanation of different regimes of friction and wear using asperity deformation models. Wear **53**, 229–243 (1979)
40. Hiroshi, M.: Surface deformation and formation of original element of wear particles in sliding friction. Wear **215**(1–2), 10–17 (1998)
41. Xie, Z., Williams, J.A.: The prediction of friction and wear when a soft surface slides against a harder rough surface. Wear **196**, 21–34 (1996)
42. Black, A.J., Kopalinsky, E.M., Oxley, P.L.B.: An investigation of the interaction of model asperities of similar hardness. Wear **153**, 245–261 (1992)
43. Koji, K., Koshi, A.: Wear mechanisms (Chapter 7). In: New Direction in Tribology, London (1997). http://home.ufam.edu.br/berti/nanomateriais/8403_PDF_CH07.pdf
44. Torrance, A.A.: The influence of surface deformation on mechanical wear. Wear **200**, 45–54 (1996)

Mechatronic System for the Promotion of Physical Activity in People with Motor Limitations

Leandro Pereira[1], José Machado[1,2(✉)], Vítor Carvalho[3,4], Filomena Soares[4], and Demétrio Matos[5]

[1] Mechanical Engineering Department, University of Minho, Campus de Azurem, Guimarães, Portugal
a75901@alunos.uminho.pt, jmachado@dem.uminho.pt
[2] MEtRICs Research Centre, University of Minho, Campus de Azurem, Guimarães, Portugal
[3] 2Ai Lab – IPCA-EST, Campus do IPCA, Barcelos, Portugal
vcarvalho@ipca.pt
[4] Algoritmi Research Centre, Industrial Electronics Department, University of Minho, Campus de Azurem, Guimarães, Portugal
fsoares@dei.uminho.pt
[5] ID+ – IPCA-ESD, Campus do IPCA, Barcelos, Portugal
dmatos@ipca.pt

Abstract. This work presents the initial step of the development of a mechatronic system that aims to promote physical activity in people with motor limitations. This mechatronic system consists of a stationary bike similar to the current gym bikes, with the difference of having a didactic component. The didactic component presents the integration of a game that simulates a paramotor in which the speed of pedalling allows the player to move up and down, while the position of the arms will allow the movement to the left and to the right in the game scenario. The model under study also presents the versatility of being able to ergonomically adjust to any person and to adjust the load for both upper and lower limbs. In this paper it is also detailed some aspects of the mechatronic system design, such as: modelling, study of materials and sensors, study of possible exercises, to be performed, and the development of the game scenario.

Keywords: Mechatronic systems design · Rehabilitation · Serious game

1 Introduction

Demographic trends, whether in developed countries or even in less developed countries, point to a sharp increase in the number of elderly people, many of them with motor problems. In Portugal, according to the Portuguese census of 2011, the percentage of young people decreased from 16% to 15% between 2001 and 2011, while the elderly population increased from 16% to 19%. Consequently, the Portuguese aging index increased from 102 to 128 between 2001 and 2011 (INE, 2012) [1]. Aging and

G. I. Gheorghe (Ed.): ICOMECYME 2018, LNNS 48, pp. 83–94, 2019.
https://doi.org/10.1007/978-3-319-96358-7_9

chronic diseases such as obesity, Parkinson's, hypertension, arthritis, diabetes and even symptoms resulting from strokes are the main factors that force people to gradually abandon their autonomy, making it more difficult to performing daily tasks such as bathing, dressing and walking. As many of the above diseases require rehabilitation, some researchers believe that active games (games involving physical activity) can increase adherence to treatment [2].

Several studies have shown that serious games have enormous potential in health intervention, including rehabilitation and physiotherapy. The pleasant atmosphere they create, and the feedback forms included, promote the interests of patients, who are increasingly motivated and involved in their rehabilitation [3]. Over the years several games have been developed with the purpose of promoting and improving the physical condition of the population in an effective and pleasant way. One of the first games of this genre was Foot Craz in 1987 and since then several games have been released, however the device that more success had was the popular Wii console of Nintendo [4]. In order to achieve the main goals of this paper, it is divided in eight sections as follows: the state of the art presents some devices that are currently used for the physical activity in an iterative way. The 3D modelling section presents the system under study as well as its modules and systems. In section selection of materials, it is made a study of the materials to be selected for the various parts of the model. In the following section it is presented an analysis to the sensors that will be part of the model making the acquisition of the inputs of the user. Then, the exercises that can be performed in the proposed model are presented in order to allow physical activity and rehabilitation. In the section development of the game scenario is presented the initial stage of the development of the game scenario, and, finally, some final remarks are presented.

2 State of the Art

In Table 1 is presented some equipment currently used for the promotion of physical activity and rehabilitation in an iterative and pleasant way.

By analysing the devices in the table above it can be verified the existence of several strengths in each of these systems, some of their strengths can be considered quite important for their acceptability and success, such as: adjustability to different people, versatility of training, dimensions, among others. Despite the great variety of interactive equipment that motivates people to exercise, it can be verified by the analysis of the table and equipment studied, that there is a need to create equipment of the same gender, but with a greater level of comfort and safety, so they can be used by people with motor limitations in a more pleasant and safe way [4].

3 Modeling of the Proposed System

The proposed 3D model presented on Fig. 1 resulted from an iterative process in which it was intended to choose the appropriated systems and to promote the desired versatility and functionality to the model. It was intended that this 3D model presented an

Table 1. Equipment currently used for the promotion of physical activity and rehabilitation

Device	Description
ReaKing	RanKinG is a serious virtual reality game that aims at the physical rehabilitation of people with mobility problems through the use of a Kinect camera [5]. This game is adjustable, transportable and compact, but it does not provide great comfort and safety
Physioland	Physioland presents a serious game developed with the objective of allowing physical therapy using a Kinect camera. The overall architecture of the game is based on five main components: the game, the hardware, the peripherals, the database and the player [6]. The Physioland is adjustable and has a structure that allows to assist the patient in the execution of movements, however the use of this structure requires some space for its installation
Smartfloor	Smartfloor is a platform game that is designed to combat childhood obesity. It was designed in a modular way, having 36 square panels of wood with load cells in each of its 4 corners [7]. This system is adjustable, transportable and compact, but it does not provide great comfort and safety to be used by people with motor limitations
Physiosensing	PhysioSensing is a pressure platform for physiotherapy and rehabilitation activities that has several functionalities, including: balance/stability, load transfer, pressure mapping and therapeutic games [8]. This equipment is compact and allows a wide variety of training, however, if the patient has motor limitations it will be necessary a person to help
Twall	Twall features an interactive wall that uses pulses of light to generate motion sequences specifically [9]. This system aims at physical and iterative training, allowing even group training, so it is not very suitable for people with motor limitations
GymTop USB	The GymTop USB is a board with visual biofeedback technology, which can be used for balance training, diagnosis of anatomical problems, resistance training and bodybuilding of the legs and torso [10]. This system is adjustable and compact, can be used in any computer and allows a good diversity of exercises and diagnoses, however, The GymTop USB does not have a great level of comfort to be used by people with motor limitations, like most of the equipment's presented in this table
BOBO Balance	The BOBO Balance is a training board that converts traditional balance devices into interactive training platforms. This technology aims to increase patient involvement, enabling specific measurements [11]. This equipment can be used in a training and rehabilitation situation, however in the case of people with motor limitations it will be necessary a person to help in performing the exercises
Trixter VR	This stationary bike offers a vast array of virtual realities, where it is possible to hold local and even international competitions [12]. This equipment is adjustable, versatile and allows the use of virtual reality glasses, however, it has the disadvantage of not being very ergonomic because it simulates a real bicycle

equipment that: occupied the minimum space as possible; had a chair with adjustment capability; allowed to adjust the loads for different workouts and to provide the necessary comfort and safety to be used by people with motor limitations.

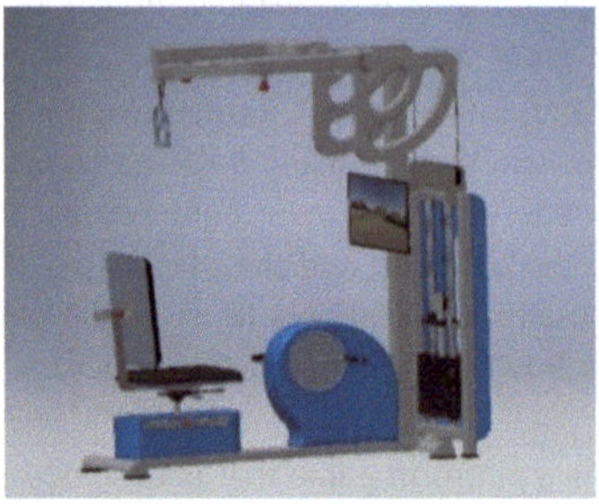

Fig. 1. Proposed solution (3D model)

To occupy the smallest space possible a modular system was chosen. In this way it will be easier to reduce the space occupied by the model when it is fully assembled. Thus, it was decided to divide the model into 3 modules (chair module, pedal module and main module) presented below. This division into modules, besides allowing the disassembly and division of the equipment, will allow the adaptation to any chair, even allowing the use of wheelchairs [4].

3.1 Chair Module

This module was the one that went through more iterative processes since it required versatility. It was intended that this module would have the following adjustment capabilities: rotary adjustment, height adjustment, pedal distance adjustment and adjustment of the backrest. In addition, this module would also aim to provide the necessary comfort and safety, so it can be used by people with motor limitations. In Fig. 2 are presented the chair module.

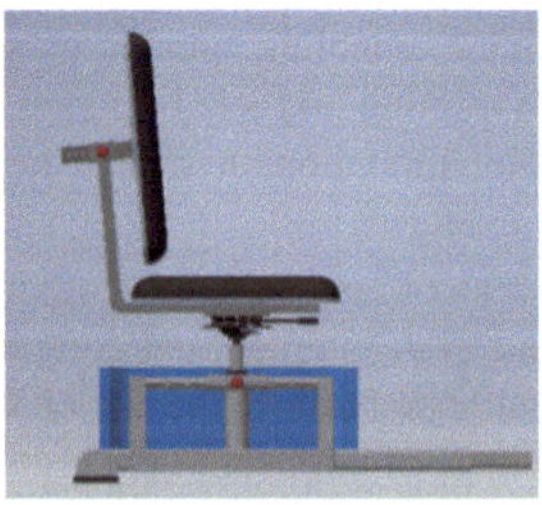

Fig. 2. Chair module

The backrest adjustment and the pedal adjustment are made by sliding between the parts of the chair and a selector pin that locks the desired position. The height adjustment is done thanks to a non-rotating pneumatic cylinder, which is embedded in the support that also allows the adjustment of the approach to the pedal. Finally, the rotation is provided with a locking swivel. Figure 3 shows the systems described.

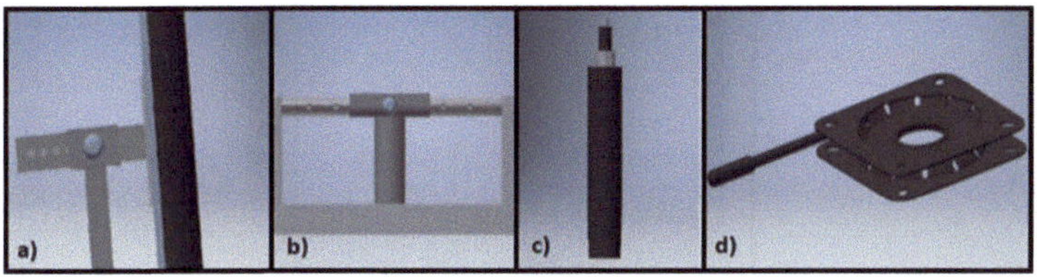

Fig. 3. (a) Backrest adjustment system; (b) pedal adjustment system; (c) height adjustment system; (d) rotation adjustment system

3.2 Pedal Module

The pedal module presented on Fig. 4 features a common system and widely used on gym bikes. This system consists of the pulley-belt assembly in which the pedalling resistance is provided by a magnetic system.

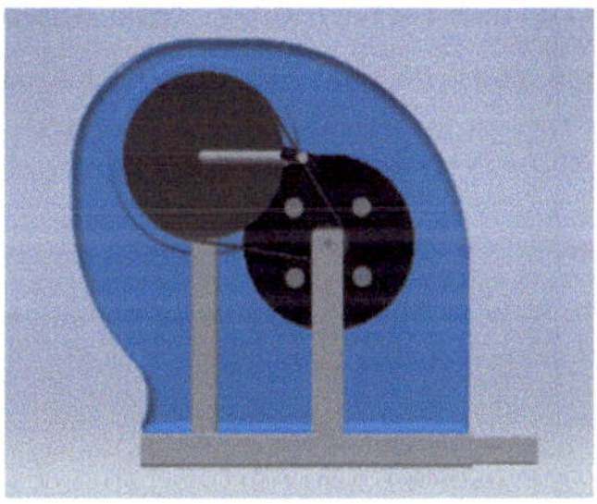

Fig. 4. Pedal module

The operation of this magnetic system shown in Fig. 5 consists of the approach and removal of magnetic supports contained inside the flywheel through a stretcher. This module should also have a speed sensor that will be discussed later in the selection of the sensors chapter.

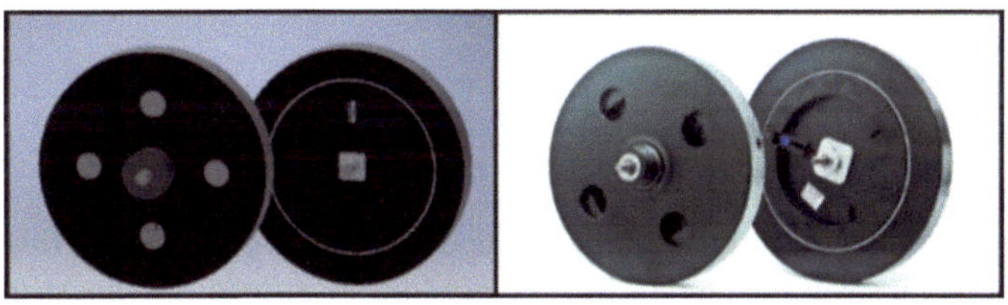

Fig. 5. 3D modeling and actual component that will be used for pedal resistance [20].

3.3 Main Module

The main module shown in Fig. 6 presents all the necessary mechanisms for the execution of the arm movements required in the game.

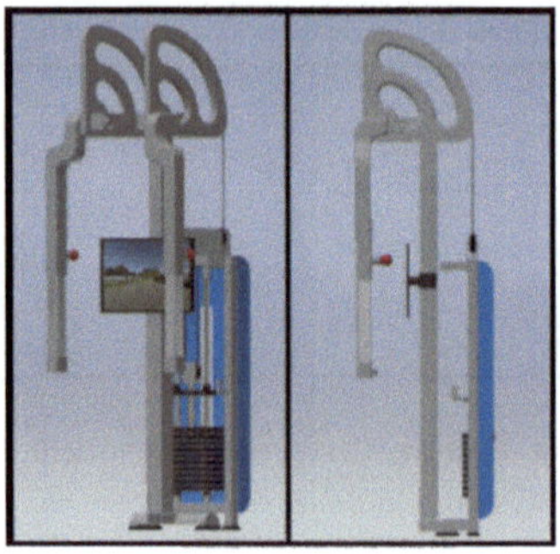

Fig. 6. Peripheral view and side view of the main module

In this module the load can be selected according to the physical capacities of the user, to do this the player must select the weight he wants from a stack of weights that are connected to the system of pulleys shown in Fig. 7. Besides the mechanisms necessary for the execution of the movements, this module also includes the game interface, and for this reason it is consider the main module. This module should also have a rotary position sensor (encoder) which will be further discussed in the selection of sensors section.

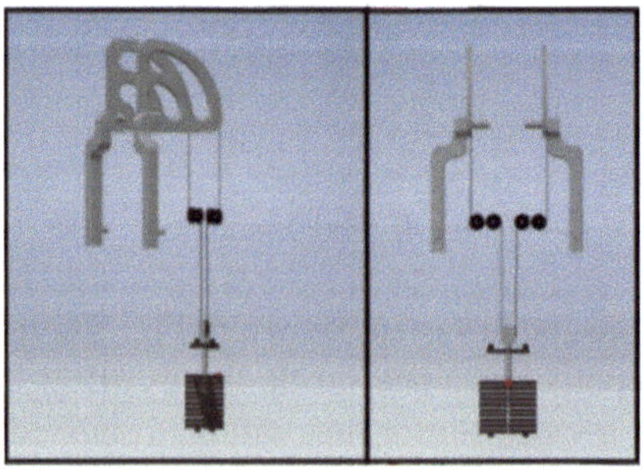

Fig. 7. Peripheral view and rear view of the pulley system

4 Selection of Materials

Bicycles and gym equipment generally use some metallic materials, mostly carbon steels, chrome steels, cast irons, aluminium and stainless steels [13]. In this context, it can be verified that several materials may be used for manufacturing the structure of the system, however the most appropriate materials, that require study, are the steels, the aluminium alloys and the polymeric materials. For the design of the structure of the proposed system, it was decided to study steels from two distinct groups, which can respond to the required needs: carbon steels (widely used in mechanical engineering) and stainless steels (important for their resistance to phenomena of corrosion).

In [14] there are presented some characteristics of a general steel; in [15] it can be found a description of stainless steels.

Table 2 shows the comparison of the different studied steels. This comparison was made based on the relative weight attributed to each of the factors previously described (mechanical strength, density, corrosion resistance, ease of processing and cost). To better understand it, a score between 0 and 5 (where 0 means that the material does not have any properties relative to the factor to which it is being evaluated and 5 means that the material has good properties relative to the factor to which it is being evaluated) was assigned to each steel for the different factors, then it was multiplied the score attributed by the relative weight of the factor and finally it was made the sum of the multiplied values in each factor.

Table 2. Comparison and selection of the materials for the structure of the model.

Factor	Relative weight (%)	AISI 1020 Score	AISI 1020 Value	AISI 1050 Score	AISI 1050 Value	AISI 304 Score	AISI 304 value	AISI 310 Score	AISI 310 Value
Mechanical strength	35	4	1.4	5	1.75	5	1.75	4	1.4
Density	5	3	0.15	3	0.15	3	0.15	3	0.15
Corrosion resistance	20	3	0.6	3	0.6	5	1	5	1
Ease of processing	10	4	0.4	4	0.4	4	0.4	4	0.4
Cost	30	5	1.5	5	1.5	3	0.9	3	0.9
Total	100		4.05		4.4		4.2		3.85

As can be seen from the table, an AISI 1050 steel was selected, but this steel does not have resistance to corrosion, in this way it is necessary to use an electrostatic powder coating treatment (coating usually used on bicycles and gym equipment) to improve corrosion resistance and increase durability

4.1 Selection of Materials to Be Used in the Adjustable Parts of the Model

The adjustable parts of the model must be resistant to corrosion without the need for coating paint, since the adjustment of the parts implies in most cases the sliding of the surfaces. In this way for the adjustable parts the selection will pass by the choice of a stainless steel (presented on the previous subchapter) or an aluminium alloy. In [16] It can be found a description of the aluminum alloys. Table 3 shows a comparison between stainless steels and aluminium alloys following the method of comparison of Table 2.

Table 3. Comparison and selection of the materials for the adjustable parts of the model

Factor	Relative weight (%)	AISI 1020 Score	AISI 1020 Value	AISI 1050 Score	AISI 1050 Value	AISI 304 Score	AISI 304 value	AISI 310 Score	AISI 310 Value
Mechanical strength	35	5	1.75	4	1.4	3	1.05	3	1.05
Density	5	3	0.15	3	0.15	5	0.25	5	0.25
Corrosion resistance	20	5	1	5	1	5	1	5	1
Ease of processing	10	4	0.4	4	0.4	3	0.3	3	0.3
Cost	30	3	0.9	3	0.9	3	0.9	3	0,9
Total	100		4.2		3.85		3.5		3.5

As can be seen from the table, an AISI 1050 steel was selected, but this steel does not have resistance to corrosion, in this way it is necessary to use an electrostatic powder coating treatment (coating usually used on bicycles and gym equipment) to improve corrosion resistance and increase durability

4.2 Polymeric Materials

For some components of the system (protections used in the modules) and for the bushes, polymeric materials will be used. Polymeric materials are widely used in the production of intermediate and finished products packaging. Nowadays most of the industries use polymeric materials in their products, in this way, it is common to find several polymeric materials in: transportation vehicles, medical equipment, scientific instruments, institutional products, furniture, electronics and clothing. The US Environmental Protection Agency states that in 2013, the United States generated about 14 million tons of plastics as containers and packaging, 12 million tons as appliances as durables-goods and nearly 7 million tons as non-durable goods such as plates and cups. There is a huge variety of polymeric materials, the most popular are: Polyethylene Terephthalate (PET), Polyvinyl Chloride (PVC), Polypropylene (PP) [17].

4.3 Electrostatic Powder Coating

Electrostatic painting is a treatment that allows to increase the resistance to corrosion and the durability of the parts. When a piece is painted with chemical powder, it receives an electric charge opposite the part, causing the powder to stick to the piece. After such a procedure, the piece is taken to an oven. When the greenhouse heats up, the paint liquefies and then hardens, forming a film of high finish, uniformity and strength.

The most widely distributed powder with the biggest market share is the Epoxy/polyester powder. It has very good chemical resistance and excellent mechanical characteristics. Its gloss durability and its color appearance are, however, mediocre. Its fields of application are industrial equipment paneling, office furniture, machine tools, water heaters, radiators and shelving [18].

5 Selection of Sensors

In order to create the game, it is necessary to acquire some inputs: detection of the pedal velocity and detection of the position of the arms of the structure. To obtain this data, it will be necessary to use encoders and sensors to measure the position and velocity. In the following section there are presented the sensors that will be used for the detection of the pedal speed and the position of the arms of the structure.

5.1 Detection of the Pedal Speed and Position of the Structure Arms

For the detection of the pedal speed it will be used a Reed Switch sensor. The Reed sensors are used as highly effective speed sensors in low speed applications (up to 1000 rpm). A Reed sensor consists of a switch having two ferromagnetic blades contained within a hermetically sealed tubular glass housing. Generally, the gas present inside the glass is the nitrogen that serves to eliminate the presence of oxygen and to ensure that the contacts do not oxidize. Reed switches are usually activated by a permanent magnet or an electromagnet. This type of sensors allows solutions for detection of proximity, speed, flow, among others. In this way these sensors are very versatile possessing different utilities and forms [19].

For the detection of the position of the arms of the structure it will be used a rotary incremental encoder. The incremental encoders are of simpler manufacture when compared to the absolute encoder. This type of encoder generates only pulses and does not have a different combination for each position. In this way, they are manufactured with a quantity of pulses per revolution [20]. The higher the number of pulses per revolution, the higher the encoder resolution. The disadvantage of this type of encoder is in the reading of its data, because as it operates by pulses, the reading of a certain positioning needs an initial reference, but they present a much lower cost when compared with the absolute encoders. Figure 8 shows the sensors that will be used.

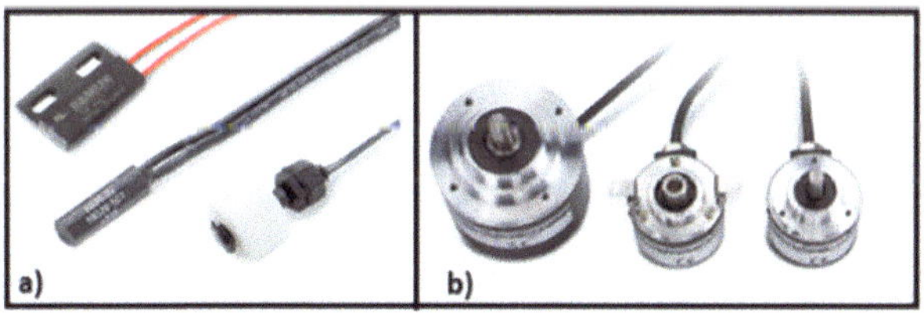

Fig. 8. (a) Reed Switch sensors [19]; (b) Rotary encoders.

6 Study of Exercises that Can Be Performed in the Proposed Model

The main exercise performed in the use of a paramotor consists of a combined movement of glen-humeral abduction/adduction (abduction consists in moving the arm away from the median plane of the body while the adduction implies the approximation to this plane) with the humeroradial flexion/extension movement (the humeroradial flexion consists of the approximation whereas the extension consists of the separation of the forearm from the shoulder, thus translating into a decrease and increase of the arm-forearm angle respectively). This movements implies the movement of the joints of the shoulder and elbow. Figure 9 shows the execution of the movements described above.

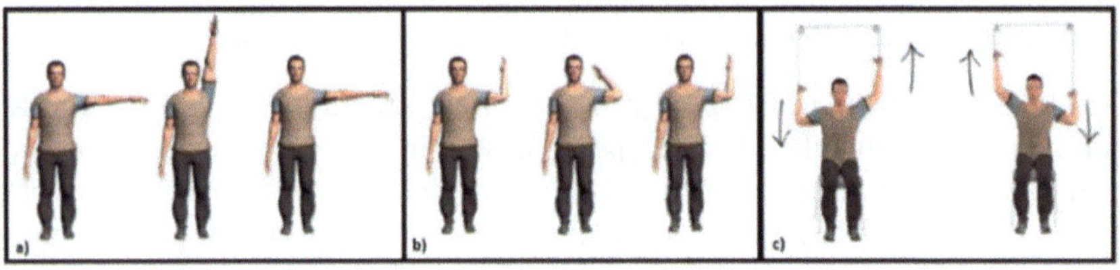

Fig. 9. (a) Glen-humeral abduction/adduction; (b) humeroradial flexion/extension; (c) combined movement of glen-humeral abduction/adduction with the humeroradial flexion/extension.

In addition to the above described exercise, other exercises, such as glen-humeral flexion/extension, may be implemented, however the exercise that best represents the paramotor conduction is the combined exercise of glen-humeral abduction/adduction with the humeroradial flexion/extension [6].

During the execution of the arm exercises, the game also features a pedal that allows the user to train and strengthen the muscles of the legs, thus making physical activity and/or rehabilitation more enjoyable. Figure 10 shows the use of the model. In this figure, it is possible to see the execution of the combined exercises of the arms and legs during the use of the game.

Fig. 10. Execution of the game exercises in the model

7 Development of the Game Scenario

The chosen game consists of the simulation of a paramotor trip in which the character starts the game when he starts to pedal. During the game the user will have to go through different regions in which they will have to deviate of obstacles that appear while picking up the objects that give points if the exercise was preformed correctly. One of the main goals will be to keep pedalling since the variation of the height will be simulated by the pedalling speed of the patient; in this case according to the speed of rotation of the pedal module, the patient will go up and down in the game scenario. Figure 11 presents some of the different regions that can be found in the game.

Fig. 11. Some different regions found in the game

8 Final Remarks

This work proposes the conceptual design of a mechatronic system that intends to promote physical activity in people with motor limitations in an iterative, safe and comfortable way. This system (as it is designed) consists in a tubular structure of medium carbon steel coated with an electrostatic paint, the movable parts shall be made of stainless steel and the shells and protections shall be of polymeric materials. The system will have a sensor to detect the speed of rotation of the pedal (Reed Switch sensor), two sensors to detect the movement of the arms (Rotary incremental encoders) and will allow the execution of the combined movement of adduction and abduction glen-humeral with the movement of flexion and humeroradial extension.

Next steps, concerning this project development will consist in the development of detailed mechatronic design and construction of the first prototype.

Acknowledgements. The authors would like to express their acknowledgments to COMPETE: POCI-01-0145-FEDER-007043 and FCT – Portuguese Foundation for science and technology within the Project Scope: UID/CEC/00319/2013.

References

1. Abreu, M., Caldevilla, N.: Attitudes toward aging in Portuguese nursing students. Procedia – Soc. Behav. Sci. **171**, 961–967 (2015)
2. Zeng, N., Pope, Z., Eun, J., Gao, Z.: A systematic review of active video games on rehabilitative outcomes among older patients. J. Sport Health Sci. **6**(1), 33–43 (2017)
3. Martins, T., Carvalho, V., Soares, F.: Web platform for serious games management. Procedia Comput. Sci. **64**, 1115–1123 (2015)
4. Pereira, L., Machado, J., Carvalho, V., Soares, F., Matos, D.: Mechatronic system for the promotion of physical activity in people with motor limitations: first insights. In: Helix 2018, 27–29 June 2018
5. Pedraza-Hueso, M., Martín-Calzón, S., Díaz-Pernas, F.J., Martínez-Zarzuela, M.: Rehabilitation using kinect-based games and virtual reality. Procedia Comput. Sci. **75**(Vare), 161–168 (2015)

6. Martins, T.: Desenvolvimento de um jogo sério para fisioterapia, monitorização e motivação de pacientes com doenças neurológicas. Tese de Doutoramento, Universidade do Minho (2018). (in Portuguese)
7. Heller, B., Senior, T., Wheat, J.: The Smartfloor: a large area force-measuring floor for investigating dynamic balance and motivating exercise. Procedia Eng. **72**, 226–231 (2014)
8. Sensing Future: Physiosensing balance and pressure plate (n.d.). https://www.physiosensing.net/. Accessed 27 Jan 2018
9. Motion Fitness: TWALL targeted training with a fun factor (n.d.). https://www.motionfitness.com/. Accessed 27 Jan 2018
10. Sensing Future: GymTop USB (n.d.). https://www.sensingfuture-store.com/. Accessed 27 Jan 2018
11. Bo&Bo Ltd.: BALANCE BETTER (n.d.). http://bobo-balance.com. Accessed 15 Mar 2018
12. Pulsefitness: New ground breaking features on Trixter VR (n.d.). http://www.trixter.net/. Accessed 28 Jan 2018
13. International stainless steel. http://www.worldstainless.org. Accessed 10 Apr 2018
14. Bramfitt, B.L., Benscoter, A.O.: Introduction to steels and cast irons. In: Metallographer's Guide, pp. 2–8. ASM International, Material Parks, Ohio (2002)
15. Davis, J.R.: Introduction to stainless steels, vol. 37, p. 577. ASM International, Material Park, Ohio (2000)
16. ASM International: Aluminum and Aluminum Alloys, p. 117. ASM International, Material Park, Ohio (2015)
17. Millholland, C.D.: A guide to the world's most widely used plastics. https://www.thermofisher.com/blog. Accessed 19 Apr 2018
18. Surface finish equioment group: The electrostatic powder coating (n.d.). http://www.sfeg.co.uk/. Accessed 20 Apr 2018
19. LittelFuse: Product catalog & design guide (sensors). http://www.littelfuse.com. Accessed 20 Apr 2018
20. Chi Hua Fitness Co.: Eddy current break (ECB) (n.d.). http://www.chihua.com.tw/en/. Accessed 30 Apr 2018

Automated System for Remote Defect Inspection

František Lopot[1,2], Daniel Hadraba[1,2], Petr Kubový[2], and Jan Hošek[1(✉)]

[1] Faculty of Mechanical Engineering, Czech Technical University in Prague, Technicka 4, Prague 6, Czech Republic
flopot@seznam.cz, {daniel.hadraba,jan.hosek}@fs.cvut.cz
[2] Faculty of Physical Education and Sport, Charles University, José Martího 31, Prague 6, Czech Republic

Abstract. The paper deals with development and application of a method which allows to perform automatic check the surface integrity of the monitored component. In our case, it is the bearing race of the rotating cylindrical container of the lime regeneration line in the paper mill factory. By visual checks, some cracks have been discovered on the circular surface of the race. These cracks have needed to be monitored to prevent fatal failure. To measure the cracks manually, it was always necessary to stop the cylinder. However, these breaks have caused momentous problems in the operation of the line, which is conceived for continuous work due to the nature of the production process. This was the main reason for the development and deployment of the automated system. Because the problem takes of low-frequency bearing run (less than 1 Hz) moreover used in outdoor conditions an image processing technology was chosen to create a tracking system to determine the immediate crack length at specified times.

Due to the very limited time span for the development of a separate robust apparatus and with regards to the fact that the system is used only till replacing the concerned race with a new one, i.e. for a period of about three months, the hardware of the apparatus was based on a standard web camera with full HD resolution connected to a notebook with a controlling and recording software created in the Matlab environment. This "recording" system located close to the monitored bearing is wirelessly connected to the "controlling" computer placed in the control center of the paper mill. The system reduces the data stream of optical scanning of 750 mm wide bearing ring by cut of the recorded figures to a data level that reliably enables their further processing and evaluation. It is done by the control center computer.

The system is able to indicate individual cracks present at the bearing ring surface with image data processing. Documentation and assessment of development of the dimensions of identified and observed cracks are also presented.

For a better idea about all aspects of the problem, we add that the ring has the diameter of 4.7 m and scanning width of almost 800 mm due to additional axial motion.

Keywords: Defect inspection · Camera · Image processing · Bearing cracks
Crack healing

G. I. Gheorghe (Ed.): ICOMECYME 2018, LNNS 48, pp. 95–103, 2019.
https://doi.org/10.1007/978-3-319-96358-7_10

1 Introduction

The collaborating paper mill factory works with very modern equipment, which allows, but at the same time supposes, non-stop operation. Any technical problems in virtually any part of the line cause irreversible degradation of raw materials in a very short time. Another problem is the fact that the stopping and start-up of some parts of the production lines, which simultaneously act as raw material trays, is a relatively long-term and technically demanding process which further increases the reported losses. Also one cannot ignore possible ecological risks.

An example of such a machine is the line for lime regeneration by a hydroxide. It is based on a mixing roller with a diameter of almost 4 m with a total length of about 40 m with rake angle of about 4° against the horizontal, which is filled up to the half of its radius at the highest point (Fig. 1).

Fig. 1. View on mixing cylinder and examined support (marked by red ellipse)

The cylinder is located in an open space with partially covered by metal sheet roofs. It is placed on several support rings, which rotate with it on two support rollers. The contact surfaces are lubricated at prescribed intervals and, due to the lubrication process, the cylinder is moved in the axial direction by ±150 mm around the nominal position. One revolution of the cylinder during normal line operation takes approx. 55 s. This time extends up to 2.5 min whenever the regular maintenance lubrication is performed. In terms of installation of the measuring apparatus, also the temperature is important – it exceeds 80 °C at the surface of the cylinder. The surface of the support ring has a temperature of about 35 °C. This difference is achieved by the design of the support ring which is conceived as a hollow with go-through holes for free air flow which effectively removes the heat emitted by the mixing cylinder (Fig. 2).

The object of the measurement shown in Fig. 1 was the support ring, on which surface cracks began to appear. The paper mill ordered a new support ring. Its delivery date was 2 months. During this time, the cracks represented a major risk of the machine failure,

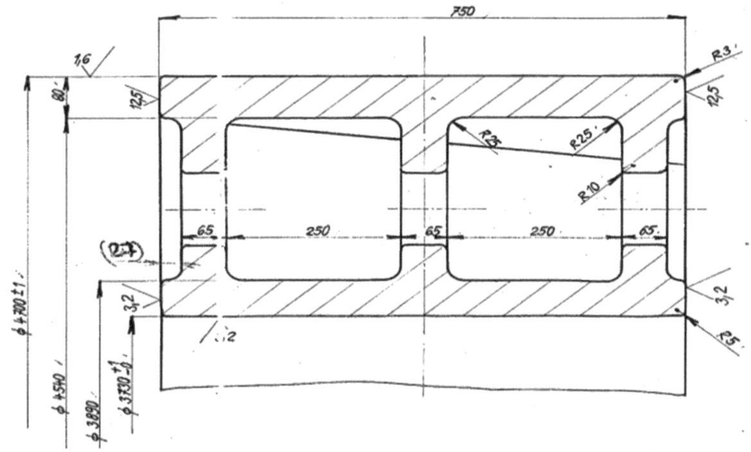

Fig. 2. Support ring profile cross section (cut-out from the drawing)

which could result in leaks of lime and hydroxide to the environment. For this reason, cracks were checked several times daily by maintenance personnel. To effectively perform that check which necessitated measurement of the actual length of the cracks identified, it was always necessary to either slow down or stop the machine/line. However, these interventions have had a negative impact on the operation of the production line as a whole, and therefore the demand for urgent creation and deployment of a reliable system for automated crack detection and identification of their immediate lengths has arisen.

To obtain such measurement system as fast as possible, starting measurement was carried out using instrumentation available at the laboratories of the CU FSPE BEL and of the CTU FME Dep. 12110. We considered to find a suitable robust solution for continuous two-month operation:

(a) Use of an ultrasound gauge
(b) Use of a linear laser 650 nm, 5 mW illumination
(c) Use of the Thorlabs LC100 1D camera (profilemeter)
(d) Image analysis (Full HD Camera, Stable LED Lighting).

The application of ultrasound gauge proved to be very problematic in the initial phase because of the dimensions of the support ring and because of the need of a contact between the ultrasound units and the measured surface. Although the projected laser beam showed changes in intensity at the cracks edges, the resulting sensitivity and accuracy of this measurement was however strongly dependent on the purity of the contact surfaces - when freshly lubricated, it stopped working completely. The recording by the 1D camera showed a lack of sensitivity and, moreover, was negatively affected by impurities on the measured surface, which could not be distinguished from the crack records. For those reasons, the final detection system was designed, based on the image analysis.

2 Method

The final measuring system was designed on the basis of a testing measurement using Dewetron equipment for measuring and controlling light conditions for determining the necessary parameters of lighting units and sensitive full HD camera with a wide set-up possibilities in terms of exposure conditions, sampling frequency and other parameters affecting the contrast and brightness of the pictures taken. In order to protect the final apparatus as much as possible from the heat and other environmental influences, it was placed under the support ring between two support rollers in the cover boxes (Fig. 3).

Fig. 3. Measurement spot installation under preliminary testing

The described testing measurement was supplemented by check of the geometric accuracy of the support ring and the supporting rollers using the Somet ČSN 25 1811 analogue device (ranged 10 mm, accuracy ±0.01 mm).

Figure 4 shows the final installation that was made up of two boxes. The camera and lighting modules - lamps were placed in the first box under the measured support ring. In the second box, the control electronics was placed – a PC with the acquisition program

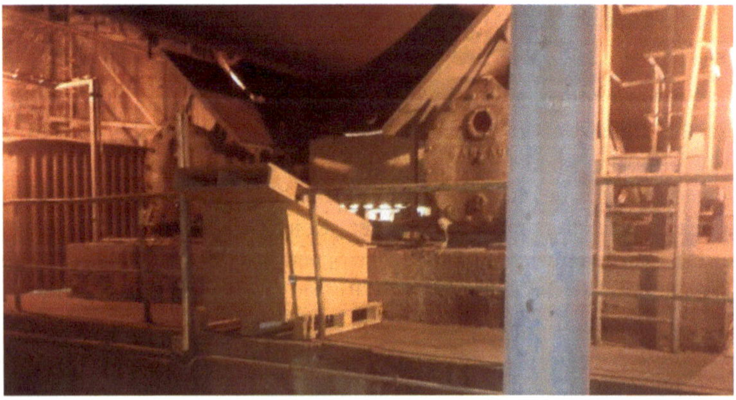

Fig. 4. Final measurement system installation

created in Matlab 2017 environment and a control unit for starting and termination the measurements based on the Arduino Uno platform. The system has recorded one revolution of the cylinder each two hours. The measurement was started and stopped by a trigger signal from a conventional optical bar located on the side of the measuring box. The protective boxes were wooden with lamellar floor to allow air flow-through for sufficient cooling of their inner space. For the same purpose, the box with the control electronics was equipped with a lamellar roof.

3 Data Processing

One revolution provided 1650 gray scale images. In order to reduce the data exchange between the control PC, cloud storage and the PC for final data processing in the control room, the reduced field of view sized of 1918 × 370 px was applied on saved images as raw data (Fig. 5a). During saving process, the images have been processed by application of smoothening filter in horizontal direction to highlight horizontally oriented objects - because the cracks occurred in a direction parallel to the axis of the mixing cylinder only. By this way, many artifacts were disposed. After that, the mathematical morphology algorithms have been applied all images were binarized by thresholding using the Otsu's method (Fig. 5b). The binary images were processed for automatic detection of cracks and their lengths evaluation. Assessed data were sent to the cloud storage from where the authorized person could download it to a computer in the control room where they could be further processed.

(a)

Fig. 5a. Raw image taken by a camera

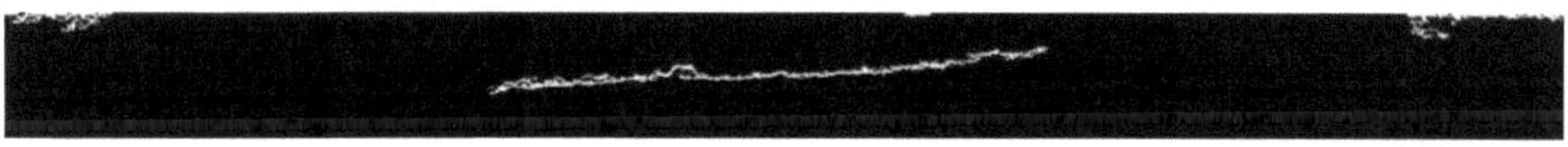

(b)

Fig. 5b. Highlighted crack after the image processing

The images prepared according to Fig. 5a were used to manually evaluate the length of the identified cracks by the authorized personnel. This evaluation was based on the output of the above mentioned automatic detection of the cracks, which identified images with suspicion of a crack. The reason for this procedure was to verify reliability and sensitivity of the automatic identification of the cracks and their lengths. The number of present cracks and two hours monitoring frequency allows to perform this manual verification one time in a working shift.

The automatic detection algorithm was based on "regionprobs" Matlab function for image processing [1]. It examines white objects in matrix of processed binary image in terms of their area and eccentricity. A found object was identified as a crack if it has minimum area of white and its eccentricity was close to one. The algorithm was extended by supplementary confirming condition requiring the repeating of the result in certain number of consecutive images. This number was determined by rotation speed of the bearing ring.

4 Results

The contact measurement of the geometry shape of the contact surfaces of the support ring and the support rollers showed strong wobbling (Fig. 6). The measurements were made at a distance of 8–10 cm from both ring faces.

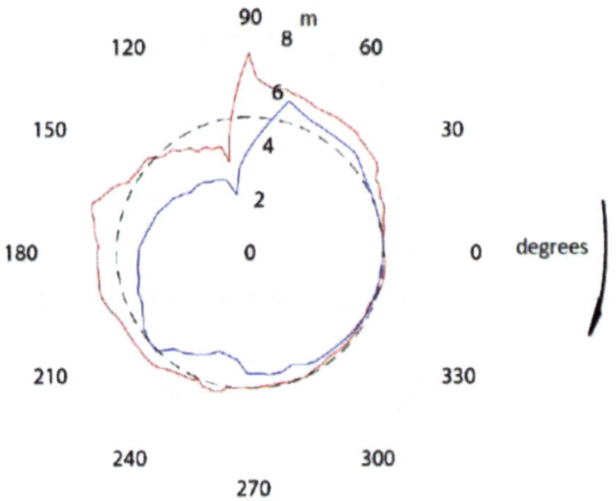

Fig. 6. Measured shape characteristics of the support ring rolling surface (red – end close to the higher face in direction of the rake angle of the mixing cylinder; blue – end close to the lower face; the scale for identified shape deviations: 1000:1)

The maximum change of the actual ring diameter was determined to 3.6 mm by repeated reading. This is 2.1 mm more than the drawing radius tolerance (Fig. 2). A step-like pattern of the measured surface curve caused induction of huge load to the bearing ring due to inertial effect of massive mill tube. The position of cracks identified by automated algorithm were localized close to the highly deformed area of the ring surface.

Ovality of the support rollers has been found in the order of max. tenths of mm without any step-wise changes.

Figure 7 presents a user interface on a computer in the control room, which was at disposal to an authorized worker of the paper mill.

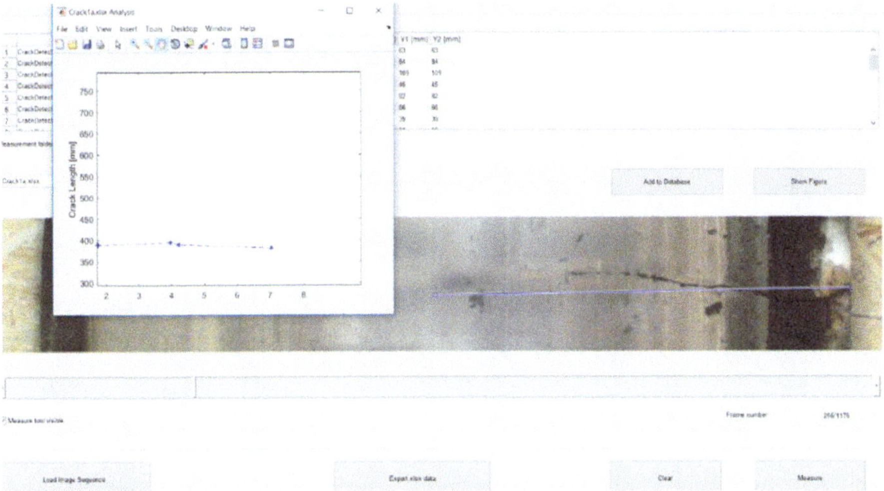

Fig. 7. The user interface of a program for final data processing and evaluation

A graphical user interface, shown in the Fig. 7, enables image upload from a cloud storage. The indicated images with suspicion of a crack were measured with a visible measuring tool. The evaluation interface contains a database in the top of the screen where the cracks are sorted according its individual measured lengths. The crack length development as a function of recorded time is generated automatically from values stored in the database in a floating window. An example of the outputs for the three largest cracks is shown in Fig. 8.

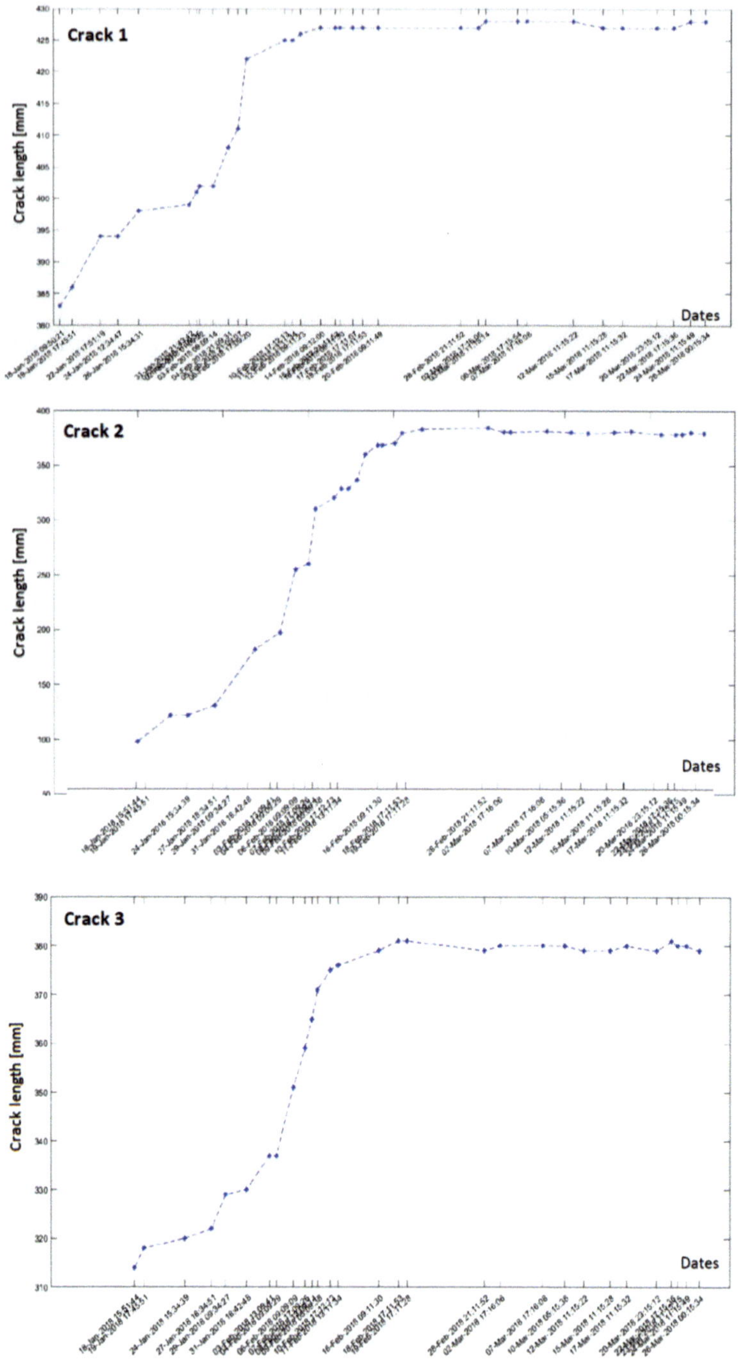

Fig. 8. Output of the system – temporal cracks length development

5 Conclusions

The aim of the paper is to highlight the importance and possibilities of continuous monitoring of the state of machine components which enables effective service operations planning. Besides that, it also facilitates service interventions because the reason of the problem is identified. Thus, the service intervention can be very effective. In early service intervention, larger damages are avoided and both service costs and its technological difficulty is also reduced. Last but not least, the on-going monitoring brings valuable data for designing and dimensioning of the monitored components.

Here, the application of the described semiautomatic system for detecting damages of the machine component eliminated the need of frequent shutdowns of the machine which was previously necessary for cracks length measurement. Online follow-up of the cracks length development further enabled quite accurate advanced estimation of the time for temporary repair of the support ring by weld rectifications of the cracks and reinforcements of the adjacent ring faces by metal plates welded in the critical places.

Surprisingly enough, pumping of about half of the cylinder content in the preparation for long-term shutdown of the production line resulted in stopping of further growth of cracks lengths, as can be seen in right parts of presented graphs. Thus, with regard to the remaining days to the exchange of the support ring, the planned temporary welding repairs could be cancelled.

Another interesting finding was that some cracks identified on the support ring in one measurement were not found in subsequent ones. This crack healing is probably attributable to the effect of rolling between the ring and support rollers [2]. It implies that manual or visual surface examination itself does not provide complete information without database of actual and former crack lengths development. For extent the system as a predictive maintenance tool, it would be beneficial to amend it by means for analyses of subsurface layers.

Acknowledgements (Facultative Field). This work was supported by project number SGS17/176/OHK2/3T/12 and 121368301825B Steti.

References

1. Mathworks web pages: https://www.mathworks.com/help/images/ref/regionprops.html
2. Gao, H., Ai, Z., Yu, H., Wu, H., Liu, X.: Analysis of internal crack healing mechanism under rolling deformation. PLoS ONE **9**(7), e101907 (2014). Kuzyk MG, ed.

Intelligent Hydraulic Power Generating Group

Mihai Avram(✉), Valerian-Emanuel Sârbu, Alina-Rodica Spânu, and Constantin Bucşan

POLITEHNICA University of Bucharest, 313 Spl. Independentei, Bucharest, Romania
mavram02@yahoo.com

Abstract. The domain of hydraulics is expensive, but this is changing due to the new manufacturing technologies. The hydraulic energy generators without an intelligent controller have low efficiency and they will be replaced by intelligent ones. The use of an inverter can significantly reduce the energy consumption by modifying the speed of the motor in order to obtain the flow/pressure required by the application. If a variable flow pump is used in the system, the energy saving is important, due to the pump working with the optimum speed, pressure and flow needed for a maximum efficiency. The use of systems connected to the Internet (and cloud services) increases the reliability due to the predictive maintenance and the reduced working costs. The paper is the starting point in developing an intelligent control system using optimized algorithms for decreasing the consumption, increasing the reliability, reducing the noise and increasing the operating safety.

Keywords: Hydraulic power supply · Intelligent system · Actuator

1 Introduction

Nowadays it is about "the fourth industrial revolution" or simply "Industry 4.0". The key of this new stage is represented by the interconnected digital systems. The manufacturing processes will become self-organized processes where human operators, machines, devices, logistics and products communicate and cooperate directly with one another. This interconnecting network must cover all the stages of the product development: concept, design, manufacture, installation, service, till recycling. In such an "intelligent factory" the products pass independently through the manufacturing processes and are easily identified and localized at any moment, giving flexibility and individualization to mass production.

Cyber-Physical Systems are complex systems capable to control and coordinate physical and organizational processes at local and global level using information and communication technology. They are the basis of control architectures for complex systems, containing autonomous subsystems as physical processes, sensors networks, communicating systems or software services. The global behaviour of such a system is given by the interaction of the components. Here are some examples of such systems: transportation systems, energetic systems, factories systems, intelligent manufacturing systems, smart-cities etc.

G. I. Gheorghe (Ed.): ICOMECYME 2018, LNNS 48, pp. 104–114, 2019.
https://doi.org/10.1007/978-3-319-96358-7_11

Two basic components are required in order to create such systems: intelligent equipment and specialists capable to develop, implement and operate the systems.

Nowadays main trends are digitization, internet of things, internet of services and cyber physical systems. Mechatronics will have an essential contribution in this domain. A mechatronic product is characterized by:

- multifunction – the capability to accomplish different tasks with one hardware structure by modifying the working program;
- intelligence – the capability to communicate with the environment and to make decisions when disturbances appear;
- flexibility – the possibility to easily modify the structure of the system in the stagers of design, manufacture or operation, due to the modular structure of the system;
- the possibility to be remote controlled and to communicate with other intelligent systems; this require complex communication interfaces;
- permanent technical evolvement, in order to meet the requirements of the changing market using the new manufacturing technologies (CNC, robots, miniaturization, nano-technologies etc.).

A system is considered to be intelligent when it has the capability to work correctly for a long period of time in uncomplete defined environments without external intervention. This requires a hardware structure that allows the system to have intelligent reactions to the disturbances: the control unit (using a microprocessor, a microcontroller, a programmable automaton or a PC) and a network of sensors providing information on the state of the mechanical structure and on the working environment. The control unit analyses the information coming from the operator and from the sensors, evaluates the state of the system and of the working environment and generates control signals to the effectors in order to carry out the programmed task. In exceptional cases, such as interferences or conflict of interests, signals are transmitted to the upper level. This in one of the four basic principles of Industry 4.0 [1].

These systems have communication interfaces and can be connected in a network with machines, devices, sensors and people, so meeting another basic principle of Industry 4.0.

2 Hydronic Systems

The development of proportional valves marked a qualitative leap in the domain of hydraulic actuating systems. Hydraulic proportional devices for controlling hydraulic power were developed using performant electro-mechanical actuators and new types of sensors. Using such devices in the structure of the hydraulic actuating systems led to the informatization of the systems and to the development of a new subdomain of mechatronics: hydronics – a synergetic joining of hydraulics, electronics and informatics. A hydronic system also contains classic hydraulic devices, sensors and transducers, electronic circuits for data processing, A/D and D/A convertors, controllers or microprocessors. Using advanced control theories, these systems can control pressure and flow rate, displacement, speed and acceleration, and also forces and torques developed by the actuators of the system. Such systems are used when automated control of the working stages is required, when speed levels shift is needed for high accuracy positioning, and when ease and flexibility of programming is required.

The designers of hydraulic control devices try to adept the other components of the system (motors, power supply groups) in order to be easily integrated in complex hydronic systems. So, a new generation of "intelligent devices" is born. This paper deals with such a system.

3 Hydraulic Power Supply Groups

The hydraulic power supply group is used to supply the flow rate of oil to the actuating system for a certain maximum working pressure. It also must maintain the temperature of the fluid within the normal limits and must assure its required purity. Usually, such a group contains the following devices: pump, tank, heat exchanger, filters etc.:

- the tank provides the fluid supply needed for the system working, protects the fluid against elements and polluting, allows the decantation of the transported particles and the separation of water and air and also contributes to the cooling of the fluid;
- the pump generates the hydraulic energy required by the system;
- the filters maintain the purity of the fluid;
- the heat exchanger maintains the temperature of the fluid under a certain limit.

The pump can be with fix or adjustable flow rate. The flow supplied by the pump is given by:

$$q = \frac{V_P \cdot n}{1000} \cdot \eta_v \left[1/\text{min} \right] \tag{1}$$

where:

V_P is the cylinder capacity of the pump $\left[\text{cm}^3\right]$;

n is the rotation speed of the actuating shaft $\left[\text{rot/min}\right]$;

η_v is the volumetric efficiency of the pump.

Usually the actuating motor (electrical or thermal) of the pump has a constant rotation speed. The pumps used in industrial hydraulic systems are actuated by asynchronous electric motors, which are robust and cheap. In such a case, the variation of the flow rate is obtained by modifying the cylinder capacity of the pump. Adjusting and stabilizing devices are commonly used: the adjusting devices require the intervention of an operator or a control signal (electric, hydraulic or pneumatic), while the stabilizing devices automatically adjust the cylinder capacity according to the flow rate, the pressure or the power according to a certain low [3].

Another method of increasing importance is the actuation of the pump with an electric motor with variable rotation speed.

The variation of the rotation speed n in the domain $\left[0, n_n\right]$ determines the variation of the flow rate q:

$$q = q_n \cdot \frac{n}{n_n} \tag{2}$$

The pressure drop on the output orifice of the pump, Δp, is given by:

$$\Delta p = \frac{1}{S_n} \cdot \frac{\rho}{2} \cdot q^2 = k \cdot q^2 \tag{3}$$

or as a function of $\frac{n}{n_n}$:

$$\Delta p = \Delta p_n \cdot \left(\frac{n}{n_n}\right)^2 \tag{4}$$

The hydraulic power supplied by the pump is given by:

$$P_h = p \cdot q \tag{5}$$

or as a function of $\frac{n}{n_n}$:

$$P_h = P_{hn} \cdot \left(\frac{n}{n_n}\right)^3 \tag{6}$$

Equations (1), (4) and (6), in a dimensionless form, are represented in Fig. 1.

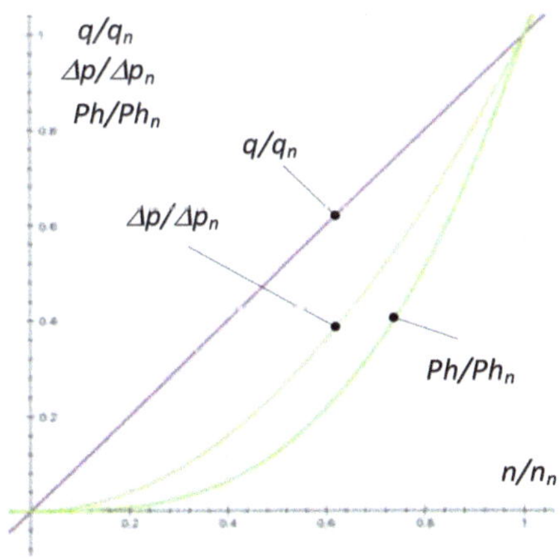

Fig. 1. Graphic representation of Eqs. (1), (4) and (6)

So, the hydraulic power for a hydraulic actuating system can be obtained using a pump of the following type [2]:

1. with fixed cylinder capacity, actuated by a constant speed motor;
2. with adjustable cylinder capacity, actuated by a constant speed motor;
3. with fixed cylinder capacity, actuated by a variable speed motor;
4. with adjustable cylinder capacity, actuated by a variable speed motor.

The type of the hydraulic power generator must be chosen after a techno-economic analysis. The cost for the life cycle of a technical system has the following structure:

- 5…20% - investing costs;
- 80…95% - operating costs: installation, training, spare parts, maintenance costs, employment costs, taxes, assurances, **energy**, modernization etc.

In the case of hydraulic actuating systems, where very high power is required, the costs with the energy are preponderant.

The following structures of a simple hydraulic actuating system can be used:

- *Actuating system with a fixed flow pump* – Fig. 2, with the following characteristics:
  - low acquisition cost;
  - low energetic efficiency; the most significant power loss ΔN_{ERC} is located at the level of the adjusting and control devices *ERC*;
  - the electric motor *M* with constant speed is coupled all the time, even when the system is not loaded, and its shaft is rotating with constant speed;
  - high power loss ΔN_P, at the level of the pump *P*, as a consequence of the high actuating rotation speed;
  - low efficiency of the motor and pump when the hydraulic actuator is partially loaded.

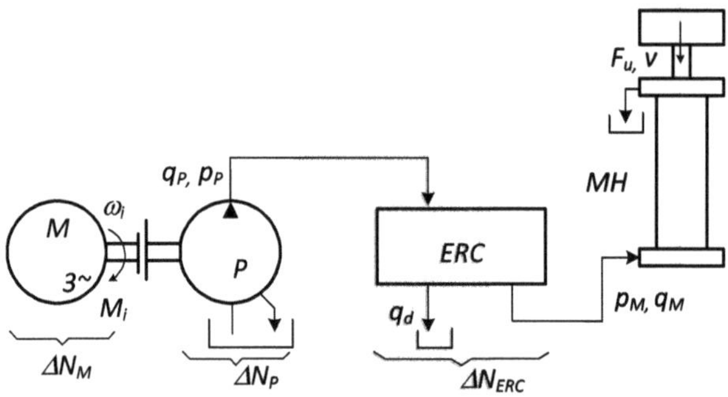

Fig. 2. Actuating system with a fixed flow pump

- *Actuating systems with adjustable cylindric capacity pump* – Fig. 3, with the following characteristics:
  - high acquisition cost, due to the complexity of the pump;
  - can be provided control devices;
  - the control devices block can be eliminated or simplified, so the power loss at this level can be significantly reduced;
  - the flow rate and the pressure at the output of the pump assures the speed and the torque needed at the shaft of the actuator.

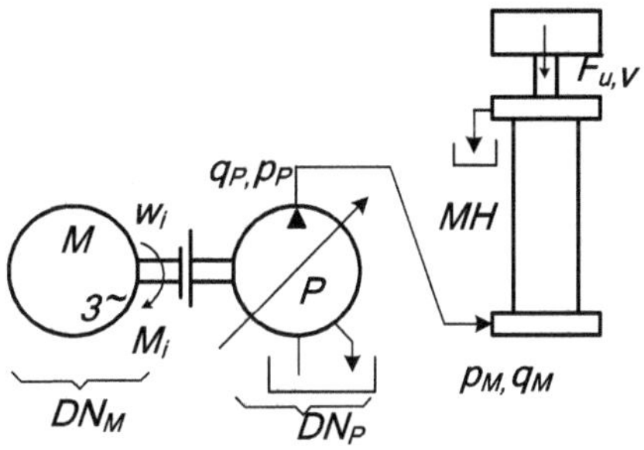

Fig. 3. Actuating systems with adjustable cylindric capacity pump

- *Actuating system with adjustable cylindric capacity pump actuated by a variable speed motor* – Fig. 4, with the following characteristics:
  - the power loss ΔN_{ERC} was eliminated;
  - the rotation speed is reduced with 40…50% when the load is partial;
  - the efficiency of the pump and the motor increases when the load of the actuator is partial;
  - the power loss at the level of the pump are reduced due to the lower actuating speed;
  - a supplementary power loss appears at the level of the invertor.

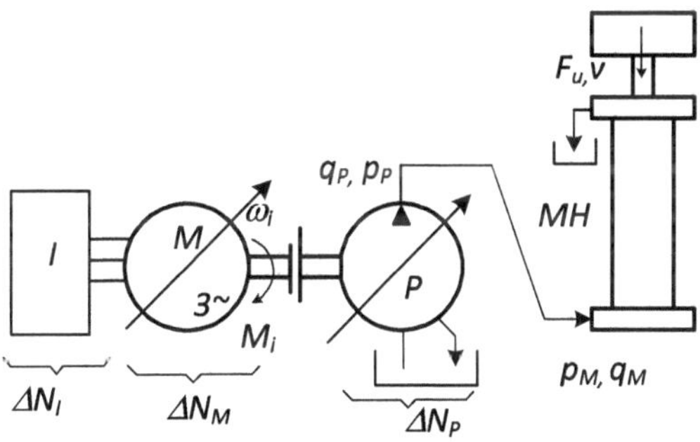

Fig. 4. Actuating system with adjustable cylindric capacity pump actuated by a variable speed motor

Nowadays, when the cost of energy is high, many manufacturers offer pumps with fixed cylindric capacity and a system for variating the speed of the actuating motor. An example is the firm Parker, who developed a pump with Denison paddles, which reduces the energetic consumption by 50%.

Another example is the firm Bosch Rexroth, with the Sytronix range [4].

4 The Proposed Intelligent Hydraulic Group

The proposed system is based on a hydraulic power generating group (Fig. 5, the blue rectangle), consisting of the pump *P* with fixed cylindric capacity, actuated by an asynchronous motor *M*, the tank *Rz* and the safety valve *Sp*.

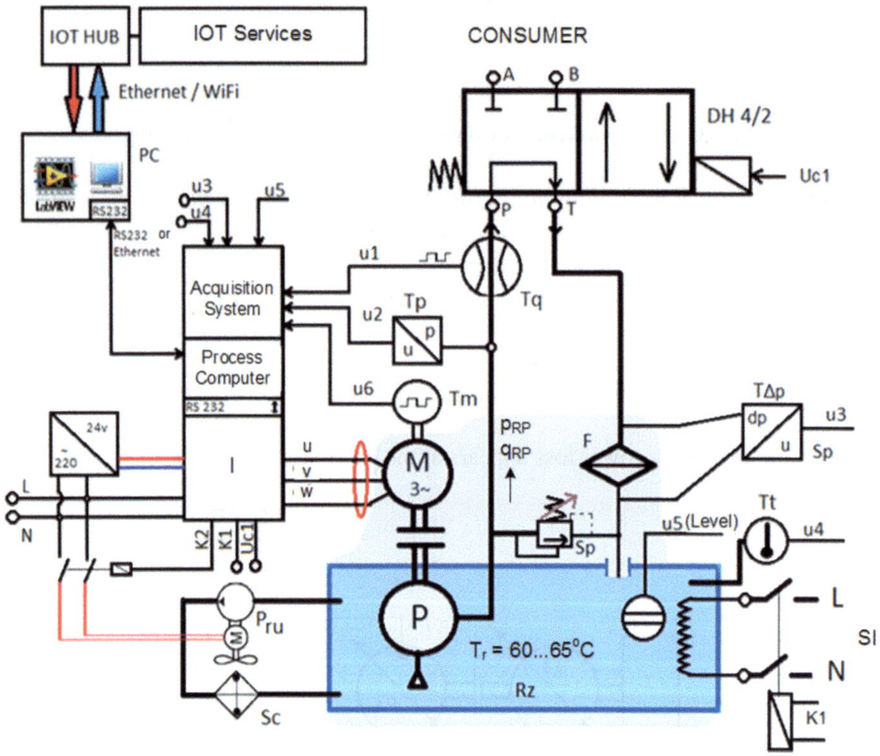

Fig. 5. The proposed intelligent hydraulic group

This structure was transformed in an intelligent one as shown in Fig. 5, and has the following characteristics:

- the supplied flow q_{RP} is according to the system needs;
- the fluid pressure p_{RP} is able to actuate the maximum load of the system;
- the fluid temperature is maintained in the regime range $t_r = 60\ldots65$ °C;
- the purity of the fluid meets the requirements of the actuated system;
- the level of the fluid in the tank can be monitored and a value under the lower limit is signalled.

The group can communicate with the control unit of the hydronic system in which was integrated.

The presented characteristics are achieved due to the following devices:

a three-phased invertor – it controls the speed of the asynchronous motor and so the flow q_{RP} supplied by the pump;

an oil heating system *SI* - it can be coupled or uncoupled by a relay K_1; this device is useful when starting the system, until the temperature of the fluid reaches the regime value;

an oil cooling system *SR* - it can be coupled or uncoupled by a relay K_2; this system consists of a fixed flow pump P_{ru} and a heat exchanger *SC* and it is coupled when the oil temperature exceeds the maximum limit of the regime range;

transducers for fluid parameters: pressure - T_p, pressure difference - $T_{\Delta P}$, flow - T_q, temperature - T_t, level - T_{ni}, rotation speed - T_n;

a 4/2 hydraulic valve with preferential position and electrical command.

The inverter *I* (Fig. 6) was designed and built with commercial electronic components, in order to lower the costs.

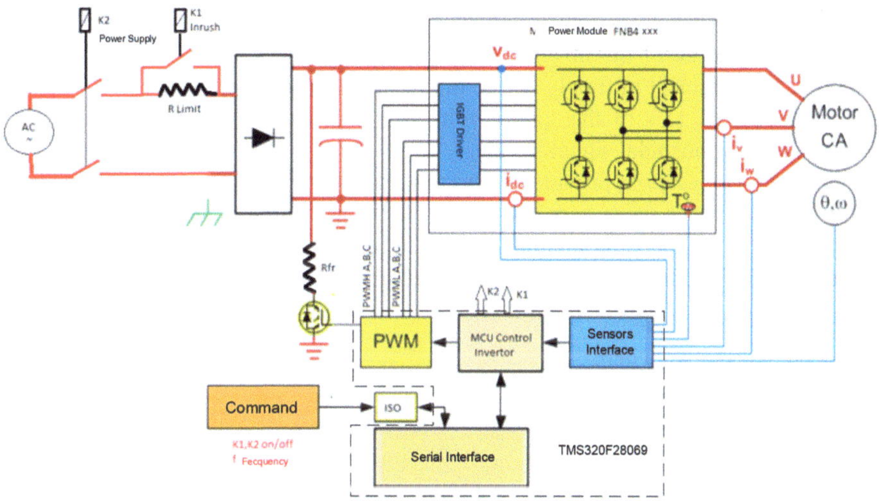

Fig. 6. The three-phased invertor

The basic component of the inverter [6] is the hybrid integrated circuit FNB43060T2 – it contains the high-power transistors and the circuits needed to activate their gates [7]. The microcontroller TMS320F28069 [8] from Texas Instruments was used to control the power transistors using 3 PWM channels with complementary output.

In order to generate the PWM signals, the method of comparing the signal to be modulated with a saw-tooth signal with a frequency of 16 kHz (or 5 kHz) was used. The modulated signal is a sinusoidal one, with the amplitude calculated as a function of the frequency f_e needed at the output of the inverter, respecting the condition that the ratio voltage-frequency is constant.

The modulated generated pulses are of low voltage, [0, 3]V, and they cannot be used to control a motor. The power module FNB43060T2 has the role to transform the PWM pulses of low voltage in PWM pulses of high voltage.

The amplitude of the fundamental harmonic [5] corresponding to the line voltage generated by the inverter depend on the dc supply voltage U_d and it can be proportionally adjusted using the amplitude modulation index m_a:

$$U_l = 0.612 \cdot m_a \cdot U_d$$

The effective value of the fundamental harmonic of the phase voltage is given by:

$$U_f = 0.353 \cdot m_a \cdot U_d$$

In Romania, the line voltage is 400 V. The DC voltage U_d is obtained using a diode rectifier and is given by:

$$U_d = \sqrt{2} \cdot 400 = 565 \text{ V}$$

These equations can be used to calculate the effective values of the line and phase voltage at the output of the inverter, for $m_a = 1$: $U_l \approx 345$ V and $U_f \approx 200$ V. These values are not sufficient to supply an AC motor in nominal conditions.

An increase of 15% of the supply voltage is obtained by injecting a third order harmonic in the sinusoidal signal [6, 11].

The proposed system is provided with a process computer – it receives the data from the transducers within the system and uses a PI/PID algorithm to generate the required command. The computer communicates with the MCU Control Inverter using an isolated serial interface at 115.2 kbaud. The recognized commands are the following:

- set the inverter frequency;
- activate the safety switch;
- activate the capacitors relay;
- read the inverter parameters (current, voltage, temperature);
- activate the brake.

Using the algorithms developed by the authors, working programmes were written and stored in the process computer memory.

The main algorithms are:

- start/stop the pump;
- automated working regime;
- safety system stop.

The process computer also assures the protection of the supplied system in the following situations:

- the motor or the pump is blocked;
- the temperature exceeds the superior limit of the regime range – the power supply generating group is deactivated;
- the working fluid quality does not correspond to the requirements - the 4/2 hydraulic valve is uncoupled;

- the oil level in the tank is not in the regime range - the power supply generating group is deactivated.

The process computer can be controlled or monitored by an upper level system. In order to connect to the internet of things, an intermediate for the protocol conversion is needed. LabView programming environment contains libraries using the MQTT protocol for communication with a IoT hub, that redirects the messages to long term storage solutions, events processing systems or other devices as phones or tablets.

5 Conclusions

The domain of hydraulics is expensive, but this is changing due to the new manufacturing technologies. The hydraulic energy generators without an intelligent controller have low efficiency and they will be replaced by intelligent ones. The control systems can be built using commercial components, leading to important energy savings only by controlling the process and managing the excess generated energy.

The use of an inverter can significantly reduce the energy consumption by modifying the speed of the motor in order to obtain the flow/pressure required by the application. Some modifications are needed (the cooling of the motor is not sufficient at low speed, the shaft must be grounded to avoid electric discharges).

If a variable flow pump is used in the system, the energy saving is important, due to the pump working with the optimum speed, pressure and flow needed for a maximum efficiency.

The use of systems that are connected to the Internet (and cloud services) increases the reliability due to the predictive maintenance and the reduced working costs. Implementing the IoT (Industry 4.0) is simple in the case of applications developed on normal computers, especially when using the graphic programming environment LabView.

The paper is the starting point in developing an intelligent control system using optimized algorithms for decreasing the consumption, increasing the reliability, reducing the noise and increasing the operating safety.

References

1. Hermann, M., Pentek, T., Otto, B.: Design principles for Industrie 4.0 scenarios. In: 2016 49th Hawaii International Conference on System Sciences (HICSS). https://doi.org/10.1109/hicss.2016.488
2. http://www.debizz.ro/bosch-utilizeaza-industria-4-0-pentru-a-si-creste-competitivitatea/
3. http://www.hydraulicspneumatics.com/sites/hydraulicspneumatics.com/files/uploads/2012/10/Komsta_Rexroth_9_11.pdf
4. https://www.boschrexroth.com/ics/cat/?id=&cat=Industrial-Hydraulics-Catalog&p=g284382
5. Albu, M.: Electronică de putere, Casa de Editură Venus
6. AC Induction Motor Control Using Constant V/Hz Principle and Space Vector PWM. http://www.ti.com/lit/an/spra284a/spra284a.pdf
7. FNB41560: Intelligent power module. http://www.onsemi.com/PowerSolutions/product.do?id=FNB41560

8. TMS320F28069 (ACTIVE) Piccolo Microcontroller. http://www.ti.com/product/TMS320F28069
9. Scalar (V/f) Control of 3-phase induction motors. http://www.ti.com/lit/an/sprabq8/sprabq8.pdf
10. Project #2 Space Vector PWM Inverter. http://www2.ece.ohio-state.edu/ems/PowerConverter/SpaceVector_PWM_Inverter.pdf
11. A Practical Guide for Connecting LabVIEW to the Industrial IoT. http://www.ni.com/white-paper/53954/en/

A Closed Form Solution for Non-linear Deflection of Non-straight Ludwick Type Beams Using Lie Symmetry Groups

M. Amin Changizi[1], Davut Erdem Sahin[2], and Ion Stiharu[3(✉)]

[1] Knowledge Engineering, Intelliquip Co., 3 W Broad Street, Bethlehem, PA 18018, USA
achangizi@intelliquip.com

[2] Department of Mechanical Engineering, Bozok University, 66200 Yozgat, Turkey
davut.sahin@bozok.edu.tr

[3] Department of Mechanical and Industrial Engineering, Concordia University, 1455 De Maisonneuve Blvd. W., Montreal, QC H3G 1M8, Canada
istih@encs.concordia.ca

Abstract. Micro Electro Mechanical Systems (MEMS) have found a large range of applications over the recent years. One of the prodigious application of micro-cantilever beams that is in use is represented by AFM probes (Atomic Force Microscopy). The AFM principle is based on the real-time measurement of the deflection of a micro-beam while following a surface profile. Hence, the prior knowledge of the deflection of beams has been of great interest to designers. Although both analytical and numerical solutions have been found for specific type of loads, there is no general solution specifically formulated for micro-cantilever beams that are not geometrically perfectly straight. Hence, the problem has not been specifically considered so far. The current work presents an analytical method based on Lie symmetry groups. The presented method produces an exact analytical solution for the deflection of Ludwick type beams subjected to any point load for non-straight beams. The Lie symmetry method is used to reduce the order of the Ordinary Differential Equation (ODE) and formulate an analytical solution of the deflection function. The result is compared with an analytical solution for a particular case that is available in the open literature. It was found that the two results coincide.

Keywords: Non-linear deflection · Ludwick material · Lie group symmetry
Micro-cantilever beams

1 Introduction

The nonlinear deflection of beams loaded by various loads and subjected to different boundary conditions has been largely investigated [1–5]. The prediction of the deflection of beams has been of great interest to researchers and designers [1–5, 8–11, 19]. This seems to be a mundane problem as it is the subject of many textbooks on elementary mechanics of materials. However, the general problem for beams that are

G. I. Gheorghe (Ed.): ICOMECYME 2018, LNNS 48, pp. 115–128, 2019.
https://doi.org/10.1007/978-3-319-96358-7_12

not geometrically straight has not been investigated so far in a systematic fashion. Moreover, materials used in applications such as microstructures exhibit deflections that are within the non-linear domain. This is one of the reasons that the Ludwick materials concept resurfaced. Materials which obey Ludwick constitutive law [1] of stress-strain are called Ludwick materials. This aspect will be further discussed further in this paper in Sect. 3. The evidence shows that micro and nano-technologies will significantly grow in the near future. This trend motivates this research. The fabrication techniques used in micro and nano fabrication may yield imperfect geometries or slightly deformed components at times. Hence, the investigation of non-straight structures represents a necessary investigation to understand the performances of slightly imperfect microstructures. This research presents an approach in solving both geometrical and material nonlinear behaviour of cantilever beams based on Lie symmetry. A general analytical solution of the problem is presented below. Hence, the objective of the present work is to investigate a versatile mathematical method that leads to a solution of the deflection of geometrically non-straight micro-cantilever beams made from Ludwick type materials and subjected to point force. The Lie symmetry method will be used for finding an analytical solution for the nonlinear ODE's. Lie symmetry is applied to the nonlinear ODE which has been formulated for large deflection of micro-cantilever beams built from a nonlinear material of Ludwick type. This ODE cannot be used for conditions for which there is no residual stress in the unloaded beam. A PDE is used to model this case.

The solution of the differential equation for the large deflection cantilever beams loaded by point forces at the tip was derived in 1945 [2]. In that approach the differential equation describing the slope of the beam versus the length of the deflected curve was formulated and solved by complete second and first kind elliptic integrals. A numerical solution was provided [3] for the differential equation of the slope versus length of the deflected beam by considering the shear force. In this work solutions were derived from finite differences method for the ODE describing a cantilever and a simple supported beam subjected to uniform distributed forces. For the simple supported beams under point forces the same method was used to solve the constitutive ODE. A numerical solution for the tapered cantilever beam under a point force at the tip was presented in 1968 [4]. A non-dimensional ODE was solved by computer at that time. The performance of a cantilever beam made from materials exhibiting nonlinear properties subjected to a point force was studied in [5].

In the results of the research presented in [6] the deflection equation was derived for a Ludwick experimental strain-stress curve. The integral equation was solved numerically to calculate the deflection and the rotation at the tip of the beam. The work in [7] used power series and neural network to solve for the large deflection of a cantilever beam under a point force at the tip. The constitutive equation that expresses the curvature of the beam made from a non-linear material is a nonlinear ODE. The authors of [8] studied large deflections of cantilever beams made from nonlinear elastic materials subjected to uniform distributed forces and a point force at the tip. They developed a system of nonlinear ODEs to model the system which was further solved by the Runge-Kutta method. The authors in [9] used the same method as in [2] to solve the large deflection equation of a cantilever under a point force at the tip. In their work they validated the theoretical work using experimental results. The non-dimensional formulation was used to simplify the nonlinear deflection and reduce the problem to linear analysis. They showed that the nonlinear assumption of small deflection yields the same results as those found through the linear analysis. Two dimensional loads on cantilever beam with point force loading at the free end was studied for both non-prismatic and prismatic beams [10]. A model for the general loading conditions on beams was formulated as a nonlinear Partial Differential Equations (PDE) [10]. Further, a numerical solution based on a non-dimensional equation was presented [10]. A theoretical study on the large deflection of a cantilever beam subjected to a tip moment made form a nonlinear bimorph material was presented using a numerical solution in [11]. The authors used an exact formulation for the deflection of a cantilever with a moment applied at the tip. The authors used a system of ODEs to numerically solve the large deflection of the cantilever beam under a uniform distributed and tip point load. They presented a comparison between numerical and experimental results. The large deflection of a non-prismatic cantilever beam subjected to different type of continuous and discontinuous loading was studied by finite difference methods [13]. The authors formulated the problem based on [10] and further used quasi-linearization central finite differences method to solve the problem. The large deflection of a cantilever beam subjected to a point load at the tip was studied by a homotopic analysis method (HAM) and an explicit solution for it was presented in [13]. Large deflection of a non-uniform spring-hinged cantilever beam under a follower point force at the tip was formulated and numerically solved [15]. Large deflection of curved beams under a tip load was studied analytically by the Lie symmetry method in [16].

In the present work the Lie group symmetry approach is used as a feasible alternative to express the curvature of a not straight beam deflecting beyond the linear level and made from a non-linear material for which stress-strain correlation follows the Ludwick formulation. Lie group symmetry is a versatile mathematical method that has recently been extended to solve some engineering problems.

2 Nomenclature

σ	Stress
E	Young's Modulcs
ε	Strain
n	Power coefficient
A	Area of cross section
I	Moment of inertia
h	Thickness of beam
V	Vertical component of force
H	Horizontal component of force
P	Point force at tip of beam
θ	Angle of point force
φ	Angle of Curvature
ρ	Curvature of beam
δ_h	Horizontal deflection of tip
δ_v	Vertical deflection of tip

y Distance from neutral axis of beam
ξ, η Components of infinitesimal transformation
U Transformation operation
α The parameter of the group

3 Ludwick Materials

The relation between stress and strain for linear materials is based on Hooke's Law which is a linear formulation [19]. A large range of nonlinear materials which exhibit nonlinear stress-strain behaviour can be defined by the Ludwick relation [5]. The Ludwick stress-strain relation is defined as following:

$$\sigma = E \times \varepsilon^{\frac{1}{n}} \tag{3.1}$$

Here, E and n represent constants related to material properties. The behaviour of this material for different n is illustrated in Fig. 1 [5].

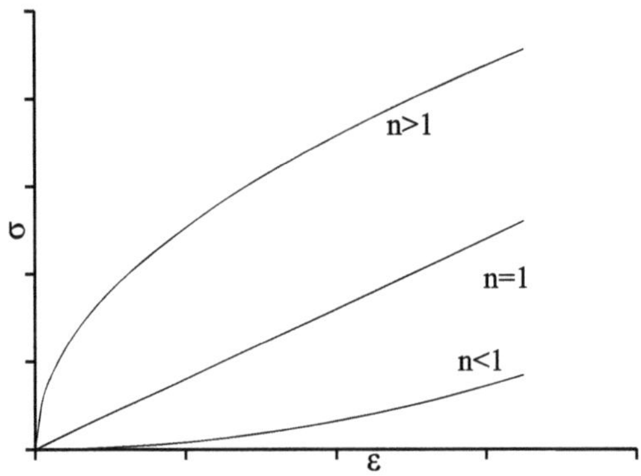

Fig. 1. Ludwick material stress-strain behaviour

In the current work the large deflection of micro-cantilever beams made from Ludwick type materials is investigated. For an example, in AFMs, certain applications require large deflection of beams which may be non-straight due to the variability of micromachining processes. This calls for a model based on Ludwick type materials as the deflection of such elements may exceed many times the initial thickness of the structure. For this case the moment of area of a rectangular cross section of the Ludwick type material needs to be defined. In the open literature has been shown that the moment of inertia can be expressed as in (3.2) [17]:

$$I_n = \int y^{\frac{n+1}{n}} dA = \left(\frac{1}{2}\right)^{\frac{n+1}{n}} \left(\frac{n}{2n+1}\right) bh^{\frac{2n+1}{n}} \tag{3.2}$$

4 Formulation of the Large Deflection of the Not-Straight Beam Problem

For a curved AFM beam, a point force at the tip can be considered as illustrated in Fig. 2 [16]. The moment at any cross section of the beam can be written as:

$$M(x, y) = V \times x + H \times y + M_0 \tag{4.1}$$

where:

$$V = P\sin(\theta)$$
$$H = P\cos(\theta)$$
$$M_0 = P.u$$

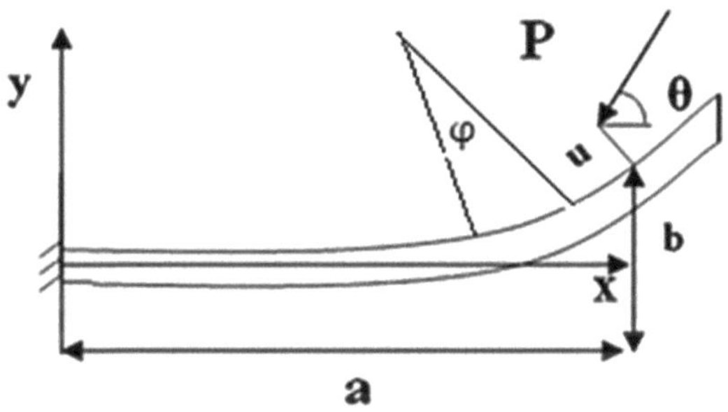

Fig. 2. A non-straight AFM micro-cantilever beam under tip load and moment.

Here u represents the distance from the tip to neutral axis of the cantilever beam. Euler–Bernoulli moment–curvature relationship results in [5]:

$$\frac{d\varphi}{d\rho} = \frac{M(x, y)}{E\left(\frac{1}{2}\right)^{\frac{n+1}{n}}\left(\frac{n}{2n+1}\right)bh^{\frac{2n+1}{n}}} \tag{4.2}$$

Where:

$$\frac{dx}{d\rho} = \cos(\varphi) \tag{4.3}$$

$$\frac{dy}{d\rho} = \sin(\varphi) \tag{4.4}$$

Differentiation of Eq. (4.2) yields:

$$\frac{d^2\varphi}{d\rho^2} = -\frac{V\cos(\varphi) + H\sin(\varphi)}{E\left(\frac{1}{2}\right)^{\frac{n+1}{n}}\left(\frac{n}{2n+1}\right)bh^{\frac{2n+1}{n}}} \tag{4.5}$$

The boundary conditions can be defined as following:

$$\varphi|_{\rho=0} = 0 \tag{4.6}$$

$$\left.\frac{d\varphi}{d\rho}\right|_{\rho=L} = \frac{M_0^n}{E\left(\frac{1}{2}\right)^{\frac{n+1}{n}}\left(\frac{n}{2n+1}\right)bh^{\frac{2n+1}{n}}} \tag{4.7}$$

where L is the length of cantilever. To solve (4.5) one should use a mathematical groups which are considered in this paper are point symmetric transformations [19] which means that each point (x, y) on the curve move into (x1, y1):

$$x_1 = \phi(x, y, \alpha) \quad y_1 = \psi(x, y, \alpha) \tag{4.8}$$

Where ϕ and ψ are diffeomorphism (C^∞). If any transformation preserves the shape of a curve and it maps the curve on itself, the transformation is called symmetry [19]. The transformations (4.8) which satisfies the group properties is called one-parameter group [19] and α is called the parameter of the group. For a one-parameter group the infinitesimal transformation is defined as [19]:

$$Uf = \xi(x, y)\frac{\partial f}{\partial x} + \eta(x, y)\frac{\partial f}{\partial y} \tag{4.9}$$

where:

$$\xi(x, y) = \left.\frac{\partial \phi}{\partial \alpha}\right|_{\alpha=0} \qquad \eta(x, y) = \left.\frac{\partial \psi}{\partial \alpha}\right|_{\alpha=0} \tag{4.10}$$

One can show that for a second order ODE infinitesimal transformation, $\eta^{(2)}$ can be calculated as in [13, 14]:

$$\begin{gathered}\eta_{\rho\rho} + (2\eta_{\rho\varphi} - \xi_{\rho\rho})\acute{\varphi} + (\eta_{\varphi\varphi} - 2\xi_{\rho\varphi})\acute{\varphi}^2 - \xi_{\varphi\varphi}\acute{\varphi}^3 + \\ (\eta_{\varphi} - 2\xi_{\rho} - 3\xi_{\varphi}\acute{\varphi})\omega = \xi\omega_{\rho} + \eta\omega_{\varphi} + \\ (\eta_{\rho} + (\eta_{\varphi} - \xi_{\rho})\acute{\varphi} - \xi_{\varphi}\acute{\varphi}^2)\omega_{\acute{\varphi}}\end{gathered} \tag{4.11}$$

where ω is right hand side of (4.5):

$$\omega = -\frac{V\cos(\varphi) + H\sin(\varphi)}{E\left(\frac{1}{2}\right)^{\frac{n+1}{n}}\left(\frac{n}{2n+1}\right)bh^{\frac{2n+1}{n}}}$$

$$\acute{\varphi} = \frac{d\varphi}{d\rho}$$

Solving Eq. (4.11) as a system of PDEs, one can calculate ξ and η. Rotation, translation and scaling as a Lie symmetry of Eq. (4.5) could be found with the following transformations:

$$\xi = C_1 + C_2\rho + C_3\varphi$$

$$\eta = C_4 + C_5\rho + C_6\varphi \tag{4.12}$$

where:

$C_1, C_2, C_3, C_4, C_5, C_6$ are constants of transformation.
Substituting Eq. (4.12) into Eq. (4.11) yields:

$$\begin{aligned}(C_6 - 2C_2 - 3C_3\dot{\varphi})\left(-\frac{V\cos(\varphi) + H\sin(\varphi)}{E\left(\frac{1}{2}\right)^{\frac{n+1}{n}}\left(\frac{n}{2n+1}\right)bh^{\frac{2n+1}{n}}}\right) \\ = \left(\frac{V\cos(\varphi) - H\sin(\theta)}{E\left(\frac{1}{2}\right)^{\frac{n+1}{n}}\left(\frac{n}{2n+1}\right)bh^{\frac{2n+1}{n}}}\right)(C_4 - C_2\rho + C_6\dot{\varphi})\end{aligned} \tag{4.13}$$

Decomposing Eq. (4.13) by comparing terms will yield:

$$C_2 = C_3 = C_4 = C_5 = C_6 = 0 \tag{4.14}$$

Hence:

$$\begin{aligned}\xi &= C_1 \\ \eta &= 0\end{aligned} \tag{4.15}$$

One can refer to Appendix B in [19] for detail calculation of constants for ξ and η Eqs. (4.15) provide a general transformation for Eq. (4.5) based on a general transformation like (4.12). Equation (4.15) presented a linear transforming along ξ-axes. One can consider ξ and η as follows:

$$\xi(\rho, \varphi) = 1 \tag{4.16}$$

$$\eta(\rho, \varphi) = 0 \tag{4.17}$$

The point symmetry operator (4.9) when using (4.16) and (4.17) will be simplified as:

$$X = \frac{\partial}{\partial \rho} \tag{4.18}$$

In the next step the attempt to solve the ODE given by Eq. (3.5) requires introduction of two variables which are called canonical coordinates, $r(\rho, \varphi)$ and $s(\rho, \varphi)$, which will be defined below. The coordinate $r(\rho, \varphi)$ is the solution

$$\frac{d\varphi}{d\rho} = \frac{\eta(\rho, \varphi)}{\xi(\rho, \varphi)} \tag{4.19}$$

and $s(\rho, \varphi)$ will be:

$$s(\rho, \varphi) = \left(\int \frac{d\rho}{\xi(\rho, \varphi(r, \rho))} \right) \bigg|_{r=r(\rho,\varphi)} \tag{4.20}$$

Substituting Eqs. (4.16) and (4.17) into Eqs. (4.19) and (4.20), respectively will yield:

$$r(\rho, \varphi) = \varphi \tag{4.21}$$

$$s(\rho, \varphi) = \rho \tag{4.22}$$

A new variable $u(r)$, can be defined as:

$$u(r) = \frac{1}{\frac{d\varphi}{d\rho}} \tag{4.23}$$

It can be shown that (for detail calculation based on Contact method, please see Appendix C in [19]):

$$\frac{du(r)}{dr} = -\frac{\frac{d^2\varphi}{d\rho^2}}{\left(\frac{d\varphi}{d\rho}\right)^2} \tag{4.24}$$

Hence,

$$\frac{d^2\varphi}{d\rho^2} = -u(r)^{-3}\frac{du(r)}{dr} \tag{4.25}$$

by replacing Eqs. (4.23) and (4.25) into Eq. (4.5) gives:

$$\frac{du(r)}{dr} = \left(\frac{V\cos(\varphi) + H\sin(\varphi)}{E\left(\frac{1}{2}\right)^{\frac{n+1}{n}} \left(\frac{n}{2n+1}\right) bh^{\frac{2n+1}{n}}} \right) u(r)^3 \tag{4.26}$$

Substitution of Eqs. (4.21) and (4.22) into Eq. (4.26) will yield:

$$\frac{du(\varphi)}{d\varphi} = \left(\frac{V\cos(\varphi) + H\sin(\varphi)}{E\left(\frac{1}{2}\right)^{\frac{n+1}{n}} \left(\frac{n}{2n+1}\right) bh^{\frac{2n+1}{n}}} \right) u(\varphi)^3 \tag{4.27}$$

One can calculate $\eta^{(1)}$ for a first order ODE like Eq. (4.27) as following [19]:

$$\eta_\varphi + (\eta_u - \xi_\varphi)\omega - \xi_u \omega^2 = \xi \omega_\varphi + \eta \omega_u \tag{4.28}$$

where ω is right hand side of (4.27):

$$\omega = \left(\frac{V\cos(\varphi) + H\sin(\varphi)}{E\left(\frac{1}{2}\right)^{\frac{n+1}{n}} \left(\frac{n}{2n+1}\right) bh^{\frac{2n+1}{n}}} \right) u(\varphi)^3$$

Substituting ω and its derivatives in Eq. (4.28) yields:

$$\begin{aligned} &\eta_\varphi + \frac{1}{E\left(\frac{1}{2}\right)^{\frac{n+1}{n}} \left(\frac{n}{2n+1}\right) bh^{\frac{2n+1}{n}}} \times (\eta_u - \xi_\varphi) \times (V\cos(\varphi) \\ &+ H\sin(\varphi)) \times u^3 \left(\frac{1}{E\left(\frac{1}{2}\right)^{\frac{n+1}{n}} \left(\frac{n}{2n+1}\right) bh^{\frac{2n+1}{n}}} \right)^2 \times \xi_u (V\cos(\varphi) \\ &+ H\sin(\varphi))^2 \times u^6 = \frac{1}{E\left(\frac{1}{2}\right)^{\frac{n+1}{n}} \left(\frac{n}{2n+1}\right) bh^{\frac{2n+1}{n}}} \times \xi \times (-V\cos(\varphi) \\ &+ H\sin(\varphi)) \times u^3 + \frac{30}{E\left(\frac{1}{2}\right)^{\frac{n+1}{n}} \left(\frac{n}{2n+1}\right) bh^{\frac{2n+1}{n}}} \times \eta \times (V\cos(\varphi) \\ &+ H\sin(\varphi)) \times u^2 \end{aligned} \tag{4.29}$$

One can show from the above equation:

$$\eta_\varphi = \xi_u = 0 \tag{4.30}$$

Comparison of the coefficients of u^3 in Eq. (4.29) will yield the equation:

$$(\eta_u - \xi_\varphi)(V\cos(\varphi) + H\sin(\varphi)) = \xi(-V\cos(\varphi) + H\sin(\varphi)) \tag{4.31}$$

To satisfy Eq. (4.29), ξ must be assumed to be zero otherwise there is no other acceptable value for it.

$$\xi = 0 \tag{4.32}$$

Hence Eq. (4.29) will be:

$$\eta_u u = 3\eta \tag{4.33}$$

The integral of this ODE is:

$$\eta = u^3 \tag{4.34}$$

So the point symmetry becomes:

$$X = u^3 \frac{\partial}{\partial u} \tag{4.35}$$

Based on (4.35), canonical coordinates can be calculated from (4.19) and (4.20) as:

$$r(\varphi, u) = \varphi \tag{4.36}$$

$$s(\varphi, u) = \frac{1}{2u^2} \tag{4.37}$$

$$\frac{ds}{dr} = \left(\frac{V\cos(r) + H\sin(r)}{E\left(\frac{1}{2}\right)^{\frac{n+1}{n}} \left(\frac{n}{2n+1}\right) bh^{\frac{2n+1}{n}}} \right) \tag{4.38}$$

Hence,

$$S = \left(\frac{V\cos(r) + H\sin(r)}{E\left(\frac{1}{2}\right)^{\frac{n+1}{n}} \left(\frac{n}{2n+1}\right) bh^{\frac{2n+1}{n}}} \right) + C_1 \tag{4.39}$$

Further by substituting (4.36), (4.37) and (4.39) in (4.27) one can get:

$$\left(\frac{d\varphi}{d\rho}\right)^2 = \left(\frac{2}{E\left(\frac{1}{2}\right)^{\frac{n+1}{n}} \left(\frac{n}{2n+1}\right) bh^{\frac{2n+1}{n}}} \right) \times (-V\cos(r) + H\sin(r)) + C_1 \tag{4.40}$$

C_1 can be calculated as following by considering boundary condition (4.6):

$$C_1 = \left(\frac{M_0^n}{E\left(\frac{1}{2}\right)^{\frac{n+1}{n}}\left(\frac{n}{2n+1}\right)bh^{\frac{2n+1}{n}}}\right)^2 - \left(\frac{2}{E\left(\frac{1}{2}\right)^{\frac{n+1}{n}}\left(\frac{n}{2n+1}\right)bh^{\frac{2n+1}{n}}}\right) \times \left(-V\cos(\theta_f) + H\sin(\theta_f)\right) \tag{4.41}$$

where θ_f is final angle in the tip. Solution of (4.40) yields:

$$\rho = \int \frac{d\varphi}{\sqrt{\left(\frac{2}{E\left(\frac{1}{2}\right)^{\frac{n+1}{n}}\left(\frac{n}{2n+1}\right)bh^{\frac{2n+1}{n}}}\right) \times \left(-V\cos(\varphi) + H\sin(\varphi)\right) + C_1}} + C_2 \tag{4.42}$$

From Eqs. (4.3) and (4.4) one can dive the following equation:

$$\int_0^a dx = \int_0^{\theta_f} \frac{d\varphi\cos(\varphi)}{\sqrt{\left(\frac{2}{E\left(\frac{1}{2}\right)^{\frac{n+1}{n}}\left(\frac{n}{2n+1}\right)bh^{\frac{2n+1}{n}}}\right) \times \left(-V\cos(\varphi) + H\sin(\varphi)\right) + C_1}} \tag{4.43}$$

$$\int_0^b dy = \int_0^{\theta_f} \frac{d\varphi\sin(\varphi)}{\sqrt{\left(\frac{2}{E\left(\frac{1}{2}\right)^{\frac{n+1}{n}}\left(\frac{n}{2n+1}\right)bh^{\frac{2n+1}{n}}}\right) \times \left(-V\cos(\varphi) + H\sin(\varphi)\right) + C_1}} \tag{4.44}$$

Which is the relationship that after the integration will yield the deflection at the tip of the not-straight cantilever beam.

5 Validation

To verify the solution of Eqs. (4.43) and (4.44), a simplified case of loading that has already been presented in the open literature [5] is analysed based on the above derived solution. The validation requires considering same conditions of loading of same type of beam that it would perform of same fashion. The assumed beam is straight as the solution in [5] is for a straight beam. In the equations the beams are described by the length (L), width (b) and thickness (h) – the geometry and the elastic modulus constant (E). The tip deflection for the two models is compared. The analytical formulation of the proposed model coincides with the solution presented in [5].

If one assumes there is no vertical and horizontal forces and there is one constant moment M_0.

Equations (4.42), (4.43) and (4.44) will be simplified as following, respectively:

$$C_1 = \left(\frac{M_0^n}{E\left(\frac{1}{2}\right)^{\frac{n+1}{n}} \left(\frac{n}{2n+1}\right) bh^{\frac{2n+1}{n}}} \right)^2 \tag{5.1}$$

$$\begin{aligned} \delta_h &= \int_0^a dx = \int_0^{\theta_f} \frac{d\varphi \cos(\varphi)}{\sqrt{C_1}} = \int_0^{\theta_f} \frac{E\left(\frac{1}{2}\right)^{\frac{n+1}{n}} \left(\frac{n}{2n+1}\right) bh^{\frac{2n+1}{n}}}{M_0^n} \cos(\varphi) d\varphi \\ &= L - \frac{E\left(\frac{1}{2}\right)^{\frac{n+1}{n}} \left(\frac{n}{2n+1}\right) bh^{\frac{2n+1}{n}}}{M_0^n} \times \sin\left(\frac{M_0^n}{E\left(\frac{1}{2}\right)^{\frac{n+1}{n}} \left(\frac{n}{2n+1}\right) bh^{\frac{2n+1}{n}}} L \right) \end{aligned} \tag{5.2}$$

$$\begin{aligned} \delta_v &= \int_0^b dy = \int_0^{\theta_f} \frac{d\varphi \sin(\varphi)}{\sqrt{C_1}} \\ &= \int_0^{\theta_f} \frac{E\left(\frac{1}{2}\right)^{\frac{n+1}{n}} \left(\frac{n}{2n+1}\right) bh^{\frac{2n+1}{n}}}{M_0^n} \times \sin(\varphi) d\varphi \\ &= \frac{E\left(\frac{1}{2}\right)^{\frac{n+1}{n}} \left(\frac{n}{2n+1}\right) bh^{\frac{2n+1}{n}}}{M_0^n} \\ &\quad \times (1 - \cos)\left(\frac{M_0^n}{E\left(\frac{1}{2}\right)^{\frac{n+1}{n}} \left(\frac{n}{2n+1}\right) bh^{\frac{2n+1}{n}}} L \right) \end{aligned} \tag{5.3}$$

These results as shown in (4.44) and (4.45) coincide with the ones provided in reference [4]. The derived solution represents a general case of the particular solution available in the open literature. The general result could be used for applications of AFMs made from nano-composites that may exhibit some curvature while the material may not have a linear behavior. However, the applications for the proposed solution are not limited to AFM). The analytical solution comes handy in the design phase.

6 Conclusion

The deflection of geometrically non-straight cantilever beams subjected point loads and made from materials behaving non-linearly such as the micro-beams used in AFM application was found analytically using Lie symmetry method which enable the reduction of the ODE describing the large deflection of the beam. Such a solution is available in the open literature for only a particular case whereas the proposed method provides a general solution for more complex geometries and loadings. Hereby general analytical solution of the deflection for the not-straight beam was methodically presented. Lie group symmetry method is a universal and powerful method for solving nonlinear differential equations.

References

1. Brojan, M., Videnic, T.: Large deflections of nonlinearly elastic non-prismatic cantilever beams made from materials obeying the generalized Ludwick constitutive law. Meccanica **44**, 733–739 (2009)
2. Bisshopp, K.E., Drucker, D.C.: Large deflection of cantilever beams. Q. Appl. Math. **3**(3), 272–275 (1945)
3. Wang, T.M., Lee, S.L., Zienkiewicz, O.C.: A numerical analysis of large deflections of beams. Int. J. Mech. Sci. **3**(3), 219–228 (1961)
4. Kemper, J.D.: Large deflections of tapered cantilever beams. Int. J. Mech. Sci. **10**(6), 469–478 (1968)
5. Lewis, G., Monasa, F.: Large deflections of cantilever beams of non-linear materials of the Ludwick type subjected to an end moment. Int. J. Non-Linear Mech. **17**(1), 1–6 (1982)
6. Monasa, F., Lewis, G.: Large deflections of point loaded cantilevers with nonlinear behaviour. Zeitschrift für Angewandte Mathematik und Physik (ZAMP) **34**(1), 124–130 (1983)
7. Ang, M.H., Wei, W., Teck-Seng, L.: On the estimation of the large deflection of a cantilever beam. In: Proceedings of the International Conference on Industrial Electronics, Control, and Instrumentation, IEEEIECON 1993 (1993)
8. Lee, K.: Large deflections of cantilever beams of non-linear elastic material under a combined loading. Int. J. Non-Linear Mech. **37**(3), 439–443 (2002)
9. Beléndez, T., Neipp, C., Beléndez, A.: Large and small deflections of a cantilever beam. Eur. J. Phys. **23**, 371 (2002)
10. Dado, M., Al-Sadder, S.: A new technique for large deflection analysis of non-prismatic cantilever beams. Mech. Res. Commun. **32**(6), 692–703 (2005)
11. Baykara, C., Guven, U., Bayer, I.: Large deflections of a cantilever beam of nonlinear bimodulus material subjected to an end moment. J. Reinf. Plast. Compos. **24**(12), 1321–1326 (2005)
12. Belendez, T., et al.: Numerical and experimental analysis of large deflections of cantilever beams under a combined load. Phys. Scr. **2005**, 61 (2005)
13. Al-Sadder, S., Al-Rawi, R.A.O.: Finite difference scheme for large-deflection analysis of non-prismatic cantilever beams subjected to different types of continuous and discontinuous loadings. Arch. Appl. Mech. **75**(8), 459–473 (2006)
14. Wang, J., Chen, J.K., Liao, S.: An explicit solution of the large deformation of a cantilever beam under point load at the free tip. J. Comput. Appl. Math. **212**(2), 320–330 (2008)
15. Shvartsman, B.S.: Large deflections of a cantilever beam subjected to a follower force. J. Sound Vib. **304**(3–5), 969–973 (2007)
16. Changizi, M.A., Stiharu, I.: Sensitivity analysis of the non-linear deflection of non-straight AFM micro-cantilever. J. Adv. Microsc. Res. **7**(1), 51–63 (2012)

17. Athisakul, C., et al.: Effect of material nonlinearity on large deflection of variable-arc-length beams subjected to uniform self-weight. Math. Probl. Eng. (2012). https://doi.org/10.1155/2012/345461
18. Changizi, M.A., Stiharu, I., Sahin, D.E.: A new approach for the non-linear analysis of the deflection of beams using lie symmetry groups. In: Proceedings of the International Conference of Mechatronics and Cyber-MixMechatronics, Bucharest, Romania, 9–11 September 2017. https://doi.org/10.1007/978-3-319-63091-5_17
19. Changizi, M.A.: Geometry and material nonlinearity effects on static and dynamics performance of MEMS. Ph.D. Thesis. Concordia University, Canada (2011)

Pipe Leakage Detection Using Humidity and Microphone Sensors – A Review

Ahmed Sachit Hashim(✉), Bogdan Grămescu, and Constantin Niţu

University POLITEHNICA of Bucharest, 313, Splaiul Independenţei, 060042 Bucharest, Romania
ahmedhashim774@yahoo.com

Abstract. In view of the recent technological development in Europe in the field of wireless power transmission systems ***WPTs***, one of the applications of these systems is the Radio Frequency identification RFIDs system. So, such a system can be used to send and receive the data, to and from sensors, to facilitate the task of detecting the leakage in pipelines, by using low-power wireless sensors, fixed like external devices and electronic devices on a mobile robot. Those sensors could be a humidity sensor for detecting oil leakage, and a microphone sensor to detect the gas leakage.

Keywords: Leak detection · Humidity sensor · Microphone sensor Mobile robot

1 Introduction

Aging and the continuous exposure to compressive stress and thermal stress in the pipelines during continuous operations, drive to the most common defects on pipes. The main reason for leakage in pipelines is the occurrence of corrosion at the first stage, the cracks occur as a second stage and then the leak, in the last stage. The reason is the poor maintenance by the oil and gas companies. The fact is that pipelines should be observed from time to time (period of maintenance) before the leak occur and cause a stop in production, preventing the risk occurrence. Therefore, for remote detection, early maintenance is used to prevent the leakage in the pipelines. The mobile robots could be a solution that can be used with various sensors. This paper presents the benefits of using a humidity sensor and a microphone sensor to provide real-time wireless capabilities and remote monitoring by mobile robots, for early detection of oil leakage or gas leakage in the external pipelines. The time and date of leakage occurrence can be determined by sudden occurrence of the gradual decrease in pressure within the pipes, using the latest communication network means and the use of appropriate sensors [1]. We can take advantage of the energy of Humidity Sensor and Microphone Sensor by using the active low-frequency range from the Radio Frequency identification RFIDs between 100 kHz to 500 kHz for its long characteristics and magnetic induction to effectively transmit and receive signals, to detect the presence of gas leak and oil leak from the coated external pipes [2]. During the development of technology in Europe, the solar transformers produced by the company Center, use transfer techniques ceramic CTTC. These

G. I. Gheorghe (Ed.): ICOMECYME 2018, LNNS 48, pp. 129–137, 2019.
https://doi.org/10.1007/978-3-319-96358-7_13

transformers work to collect the solar energy to convert into DC to produce active frequencies for their radiation to be used in WPT. The WPT is one of the applications of RFIDs. So, the RFIDs can be used to send and receive the data, to and from sensors, to facilitate the task of detecting the leakage, by using low-power wireless sensors, fixed like external devices and electronic devices on a mobile robot. The sensors could be a humidity sensor for detecting oil leakage, and a microphone sensor to detect gas leakage on the external pipelines [3]. Experimentally, the location and size of the leakage in a gas pipeline using the microphone was detected with a significant increase in noise which is caused by the leakage. It was observed using the artificial neural network, depending on Radial Basis Function (RBF) Network (one of the artificial neural network application) [4]. The artificial signal and detection by using a microphone could be simulated, as a result of the pressure balance generated in the pipe by friction between the walls in the pipe. It is also the best way to detect leakage in gas pipelines [5]. The wireless system to detect the external leakage of gas and oil in the pipes using a humidity sensor and a microphone sensor was used in some applications. It is based on an Arduino microcontroller and it is a non-destructive technology (NDT). The data is obtained from the sensors, then transferred via ZigBee and then processed by a powerful software (for example LabVIEW). If it is necessary, the results can be transferred to a Web page [6]. Artificial neural network technology could be used, to detect gas leakage in pipelines using a microphone sensor to analyze the sounds of noise dynamics using different frequencies as input into the neural model. The size and location of the leakage from the model outputs were determined, and the results showed that method gives the detection accuracy 100% on the leakage [7]. Vibration waves that pass along the walls of pipes resulting from leaks was detected. After that, using FFT as a mathematical tool, it was calculated the dominant peak on the frequencies, where the frequencies range was about 20–200 Hz [8]. A wireless sensor network system WSN has been used to detect gas. The effectiveness of this system, its accuracy and speed in response to gas leakage, and the monitoring and assistance of the Environmental Meteorological Service have been demonstrated to detect the risks of gas diffusion in the atmosphere as well [9].

2 Detect Leakage in Gas and Oil Pipelines

Leakage is defined as a method of escape or the exit of the gas and the oil due to a-holes occurrence, due to exceeding the maximum permissible for overwork of pressure and temperature inside the pipelines. This type of behavior is undesirable, and it is a dangerous defect, causing significant losses in physical material (and economic losses). Leakage could also drive to enormous environmental disasters. Safe, continuous detection and adequate protection of pipes should be provided to prevent leakage to occur. Helped by the evolution of modern technology in the last two decades, the leakages in the pipes can be detected by using some remote sensors, to determine the location and size of the leakages. There are four important activities to reduce the occurrence of leakages in the gas and oil pipelines as follows: pressure management within the pipes; effective acceleration control; speed and quality repairs in the production; maintenance, repair, and replacement of damaged pipelines with regular periodic times.

3 The Functional Description of Wireless Power Transmission Systems WPTs with Radio Frequency Identification System RFIDs

Transmitter system is a device that generates a constant wave signal CW at 2.4 GHz, amplifies the RF signal, and then the amplified signal is emitted by an antenna.

Receiver system - the RF signal is received by an antenna, where the rectifier system converts the receiving signal 2.4 GHz to a DC voltage signal. Of course it is necessary to have a battery in order to power the circuit with suitable voltages to enable it to accomplish the task.

4 Benefits of Wireless Power Transmission Systems WPTs

The advantages of wireless power transmission systems are based on the maximum RF efficiency to transmit, receive, amplify and convert DC signals as follows and as shown in Fig. 1.

- Increased energy transfer density.
- Generate and amplify the frequency signal and its radiation by an antenna.
- Generated and control the volume of energy required for the circuit.
- Reduce the range of harmonics.
- Specify the total size of antenna [10].

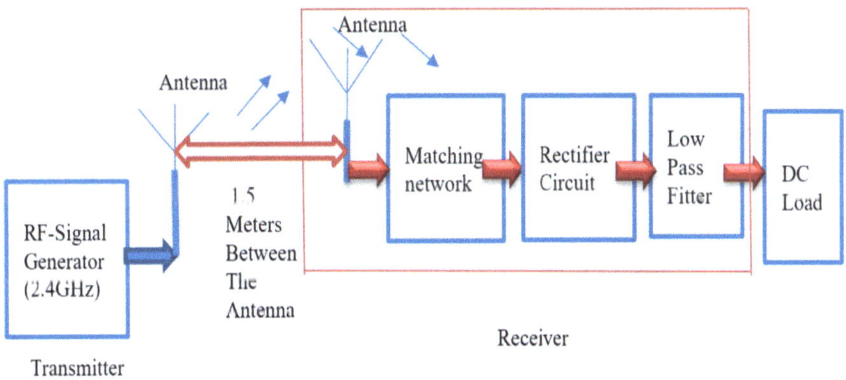

Fig. 1. The working WPTs with the RFIDs [10]

5 Why the RFIDs System Was Selected for Detection Purposes?

RFIDs have been chosen because it has long characteristics and magnetic induction is good. The RF signal can be radiated effectively if the antenna dimensions are similar to the operating frequency wavelength. Where the frequencies are very low between 100 kHz to 500 kHz, and because the wavelength of the operating frequency is in kilometers, therefore, the antenna must be well designed and installed with pre-calculated distances on the pipes insulators.

For example, in North America, antenna transmission capacity was 36 dBm, where most RFID tags require microchips of 10 dBm or better for the purpose of obtaining good and accurate signal response. So, the low frequency system RFID could operate with a humidity sensor and microphone sensor, and it must match the high-precision ideal resistance between the RFID chip and tag used in the humidity sensor, for example. The signal from the RFID system is proportional to the relative humidity resulting from detecting on crude oil leakage on the external pipes.

6 DHT22 Humidity Sensor

Moisture is an unwanted contaminant and is capable of penetrating virtually any surface, including metals such as copper, aluminum, bronze, and carbon steel. Therefore, it is important first to accurately measure moisture content, to control or remove unwanted moisture. By understanding the time and how to measure and manage humidity, we are able to improve product quality, reduce equipment damage, save energy, reduce costs, and meet the obligations to provide protection for pipes in particular. The negative side of moisture includes condensation, corrosion, cracking, leakage, pollution and many hazards. Obviously, moisture has the potential to cause expensive problems (Fig. 2).

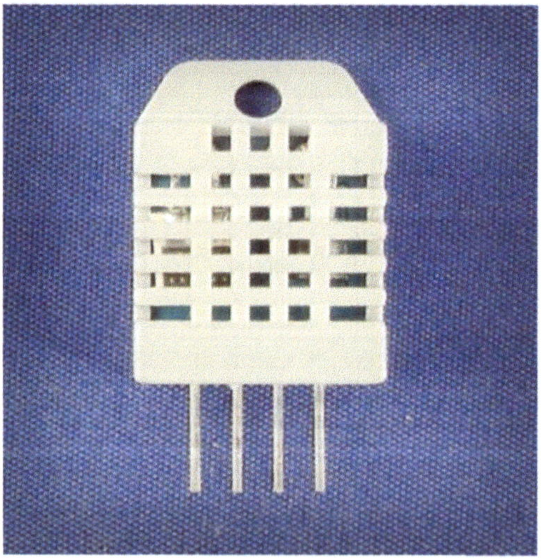

Fig. 2. DHT22 humidity sensor [11].

Description of the DHT22 moisture sensor - digital signal with calibration output. It uses an exclusive technology for digital signal collection and moisture sensor technology, ensuring its reliability and stability. Its sensor elements are connected to an 8-bit mono chip; they are installed on a computer or on mobile robots. Small size, low power consumption and long-distance transmission (20 m) made DHT22 suitable and easy to use.

DHT22 Moisture Sensor Operation Specifications:

- Energy processed

The voltage must be 3.3–6 V DC. When power is supplied to the sensor, do not send any instructions to the sensor within one second to pass the unstable situation. One 100 nF capacitor can be added between VDD and GND to filter the waves, as shown below in Table 1.

Table 1. The technical specification for the DHT22 moisture sensor [11].

Power supply	3.3–6 V DC
Output signal	digital signal via single-bus
Sensing element	Polymer capacitor
Operating range	humidity 0–100% RH; temperature – 40–80 °C
Accuracy	humidity ±2% RH (Max ±5% RH); temperature <±0.5 °C
Resolution or sensitivity	humidity 0.1% RH; temperature 0.1 °C
Repeatability	humidity ±1% RH; temperature ±0.2 °C
Humidity hysteresis	±0.3% RH
Long-term stability	±0.5% RH/year
Sensing period	Average: 2 s
Dimensions	small 14 mm × 18 mm × 5.5 mm; big 22 mm × 28 mm × 5 mm

- The system of Communications

Data is used to communicate between MCU and DHT22, costing 5mS for one-time communication. The data consists of the integral and decimal part. That is, the DHT22 sensor sends the data slightly higher or lower, that is, DATA = Integrated RH data 8 bits + 8 Decimal RH data + 8 bits Integrated T data + 8 bits T + 8 bits Optional value If data transfer is appropriate, the total checksum is the last 8 bits of 8 data integrated RH + 8 bits decimal data decimal RH + 8 bits integrated T data + 8 decimal bits. When the MCU sends the start signal, the DHT22 sensor changes from the low power consumption mode to the power-on mode. When the MCU finishes sending the start signal, the DHT22 sensor sends a response signal from 40-bit data that reflects the occurrence of relative humidity, as shown below in Table 2.

Table 2. The electrical characteristics of the DHT22 moisture sensor [11].

Item	Condition	Min.	Typical	Max.	Unit
Power supply	DC	3.3	5	6	V
Current supply	Measuring	1		1.5	mA
	Stand-by	40	Null	50	μA
Collecting period	Second		2		Second

- Working conditions

The DHT22 sensor works in a good and stable condition when the temperature is between 50 °C and 60 °C, humidity <10% RH for two hours. The DHT22 sensor works in a good and stable condition when the temperature is between 20 and 30 °C, humidity >70% RH for 5 h. Relative humidity is strongly dependent on temperature, which is why we use temperature compensation technology to ensure RH measurement accuracy. Therefore, we should focus on the work of the sensor DHT22 humidity with the recommended temperature and humidity as above, as shown below in Fig. 3.

Fig. 3. The reading of moisture and temperature

The following graph of the DHT22 Humidity Sensor reads the relative humidity and the temperature (°C), as can be observed in Fig. 4,a, and Fig. 4,b, respectively.

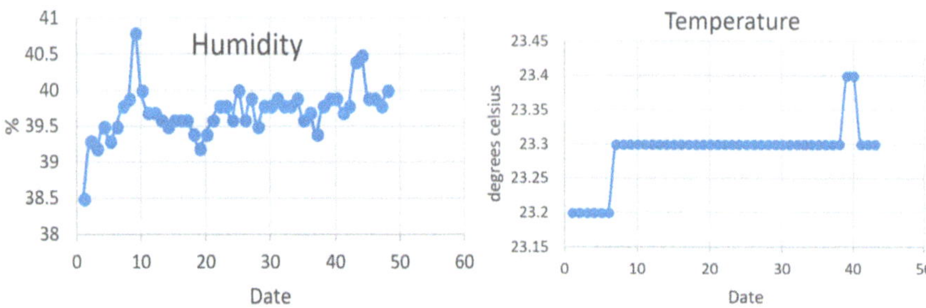

Fig. 4. DHT22 humidity sensor reads the humidity (left); b. the temperature (right)

7 MQ-6 Gas Microphone Sensor

The microphone sensor detects and monitors the leakage resulting from gas mass flow at high speed and high pressure, because of the high sound and pipe vibration occurrence due to the leakage. Pipeline leakage, lead to a reduction the pressure inside the pipes, loss, and degradation of products, environmental pollution, and many risks.

The technical data for MQ-6 gas microphone Sensor is as follows:

- the features

It has high sensitivity to LPG, ISO, butane, and propane. Small sensitivity to alcohol and smoke. Fast response. A stable and long life; Circuit design in a good mode with a simple Specifications.

- operation and control detection sensitivity with the MQ-6 gas microphone sensor

The value of MQ-6 resistance is affected by different types of gases and their different concentrations. Therefore, when using this device, the sensitivity must be adjusted and calibrated when it detects 1000 ppm of the LPG concentration in its environment. The load resistance value (RL) used is about 20 K (10 K to 47 K). When measuring accurately, the appropriate warning point for the gas detector should be determined by considering the effect of temperature and humidity as shown below in Fig. 5.

Fig. 5. MQ-6 microphone sensor for gas.

8 Mobile Robot for Leakage Detection

For the leakage detection it is necessary to have a mobile support for sensors. Next rows present the external pipe locomotion of a robot for inspection and monitoring of the pipelines for use in oil and gas industry (Fig. 6). During regular motion, the robot resembles to a crank-slider mechanism with equal lengths of the crank and rod, while for stepping over flanges it becomes an open chain with double actuated leverage. It could be used for inspection and monitoring purposes, detecting the most common defects on the horizontal and vertical external pipes. It needs to be able for receiving and transmitting data to and from the worker.

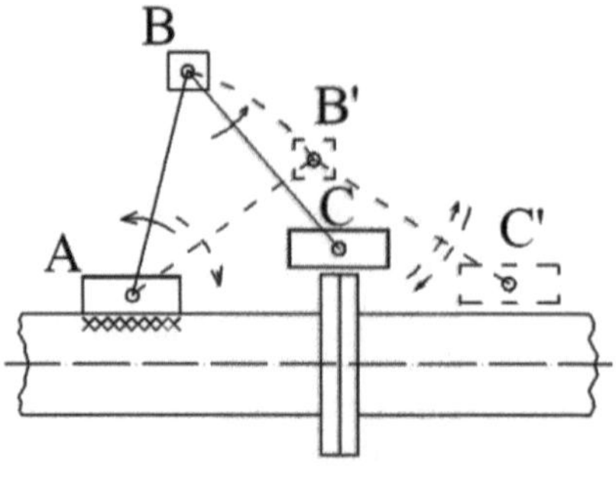

Fig. 6. Mobile robot locomotion and jumping on the obstacle to detection.

Mobile robot has the ability to move along the pipes, to negotiate the curves of the pipes (Fig. 7).

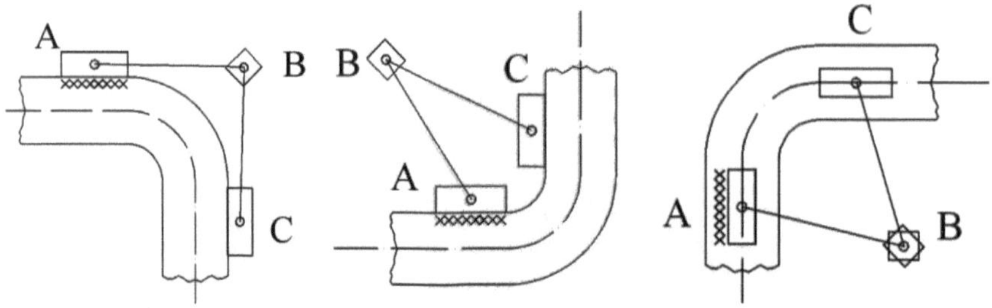

Fig. 7. Robot at turns down and up; robot at turns right and left

9 Conclusion

WPTs using radio frequency identification (RFID) systems are efficient, reliable and provide high-resolution. Using such a system, a mobile robot could have success in detecting the most common defects above the external pipes. The robot should have the ability to carry and install the external devices and electronic devices on his body. The humidity sensor has been used to detect the crude oil leakage and the microphone sensor to detect the leakage of gas on the external pipelines.

References

1. Ejofodomi, O., Ofualagba, G.: Exploring the feasibility of robotic pipeline surveillance for detecting crude oil spills in the Niger delta. New Brunswick, Canada, IJUSEng **5**(3):38–52 (2017)
2. Nasir, A., Soong, B.H., Qaraqe, K.A.: RFID in-pipe moisture sensing system for oil and gas quality monitoring in Qatar. Texas A&M University at Qatar and Nanyang Technological University (2013)

3. Carvalho, N.B., Georgiadis, A., et al.: Wireless power transmission: R&D activities within Europe. IEEE Trans. Microw. Theor. Tech. **62**(4), 1031–1045 (2014)
4. Santos, R.B., Ruppb, M., Santiago, J., Bonzi, S.J., Filetia, A.M.F.: Comparison between multilayer feedforward neural networks and a radial basis function network to detect and locate leaks in pipelines transporting gas. In: The Technical University of Berlin in Germany and the University of Campinas in Brazil, Chemical Engineering-Transactions, vol. 32 (2013)
5. Adnan, N.F., Ghazali, M.F., Amin, M.M., Hamat, A.M.A.: Leak Detection in Gas Pipeline by Acoustic and Signal Processing – A Review. IOP Publishing (ICMER), Malaysia (2015)
6. Adsul, S., Sharma, A.K., Mevekari, R.G.: Development of leakage detection system. In: International Conference on Automatic Control and Dynamic Optimization Techniques (ICACDOT), International Institute of Information Technology (I^2IT), Pune, India (2016)
7. Santos, R.B., de Sousa, E.O., da Silva, F.V., da Cruz, S.L., Fileti, A.M.F.: Real-time Monitoring of Gas Pipeline Through Artificial Neural Networks. School of Chemical Engineering, University of Campinas, Brazil (2013)
8. Karkulali, P., Himanshu, M.: Leak detection in gas distribution pipelines using acoustic impact monitoring. In: Energy Research Institute and School of Electrical & Electronics Engineering in Nanyang Technological University, Singapore (2016)
9. Luo, Q., Xiaoran Guo, X., Wang, Y., Xufeng Wei, X.: Design of Wireless Monitoring System for Gas Emergency Repairing. University of Beihang, Beijing, China, IEEE (2016)
10. Shastry, P., Baliss, E., Sanchez, S., Hoge, T.: Wireless power transfer system WPTs. Department of Electrical and Computer Engineering, Bradley University in Peoria (Illinois), USA (2013)
11. https://www.sparkfun.com/datasheets/Sensors/Temperature/DHT22.pdf

Current State of Anthropometric Parameters Measurement with Relevance for Advanced Lens Optometric Compensation

George Baboianu(✉), Constantin Nitu, and Constantin Daniel Comeaga

University POLITEHNICA of Bucharest, Bucharest, Romania
geobaboianu73@gmail.com

Abstract. In recent years, in optometry, more emphasis is laid on the accurate measurements of the individual parameters depending on the main field activity of a patient during a working day. The lens built with individual parameters is a lens that gives the wearer maximum comfort. The devices for measuring these parameters have changed considerably, from an ordinary ruler for measuring the pupillary distance, to the latest generation devices that, by taking pictures, calculate all the needed parameters. The devices for parameters measuring are mobile or stationary units and include complex software that converts pixels that make up the picture took to the patient in distances. Hence, measurements can be entered into the software generator Free-Form of lens processing to customize them. The article presents the common problems in measurement of anthropometric parameters necessary for an advanced optometric compensation. The measurement methods, from very simple and cheap to actual image-processing based method, are presented with an emphasis on disadvantages and advantages of each one. Future direction of improvement is included.

Keywords: Optometric compensation · Anthropometric measurements
Free-form lens · Image-processing

1 Optometric Parameters – Patient's Individual Parameters

Ophthalmic optometry is the science that deals with the investigation of the functioning of the human visual system, analyzes the results, highlighting the problems, recommending methods and means of seeing improvement, in order to achieve visual comfort in relation to the needs of the subject, but without medical treatment, if possible [1]. Investigating the functioning of the visual system, analyses of the results, highlighting the problems, recommending the methods and means of improving the functioning, in order to achieve visual comfort in relation to the needs of the subject represents the science called Ophthalmic Optometry.

In recent years, optometry has developed more, considering the individual as an integral part of the living environment, analyzing the performance and visual problems daily, depending on the work developed during a day. The maximum visual performance achieved by a patient is due to the measuring methods and means of the optometric parameters. In recent years more emphasis is laid on the accurate measurements of the

G. I. Gheorghe (Ed.): ICOMECYME 2018, LNNS 48, pp. 138–148, 2019.
https://doi.org/10.1007/978-3-319-96358-7_14

individual parameters depending on the main field activity of a patient during a working day. Since people have facial asymmetry, wearing a glasses frame is not exactly easy. There are people who have a deviated septum or differences in positioning of the ears and even a different horizontal and vertical positioning of the eyes. All these asymmetries that are commonly found in the patients' physiognomy must be taken into account in order to build a proper pair of glasses fit to the needs.

Parameters as inter-pupillary distance, vertex distance, convergence, are the parameters to be inserted into the lens calculation for a maximum comfort of the wearer. Also the chosen glasses frame and parameters are very important, now the new optical system consisting of the eyes and frame that holds the lens according to the prescription. The frame basic parameters are the pantoscopic angle and the curvature angle which, along with the constructive dimensions, represent an integrant part of the calculus formulas for a customized lens.

A precise optometric compensation based on modern free-form lenses, optimally designed and manufactured for each patient require an exactly evaluation of visual system and face dimensions.

1.1 Pupillary Distance

The pupillary distance or the interpupillary distance is the distance measured in millimeters between the centers of the pupils and it is used in glasses prescription. Positioning the lenses correctly, according to the pupil center is vital to achieve optimal results in vision correction or for the improvement of the visual quality.

The pupillary distance value varies from one person to another taking into account gender, ethnicity, age, region and other values. There is a difference between men and women and between US and European or between Asia and America. Thus, for people in Europe, where Romania falls in, the normal value for adults is between 53 and 67 mm and for children between 40 and 54 mm. These are the most common values which can be within normal limits, however the most common values among men in proportions of over 50% is of 69 mm and of 64 mm among women [2].

Although the pupil is a circular hole located in the center of the iris diaphragm of anyone's eyes, its diameter normally varies between 2 to 5 mm, according to the age (higher in children, lower in elders) and to the brightness to which the eye is exposed to, or in the case of certain functional states of the eye. It was observed that in 80% of people, the left and right intervals are not symmetrical, the difference reaching up to 5 mm (generally the left side is lower).

1.2 Vertex BVD Distance (Back Vertex Distance)

Vertex distance, the distance between the cornea and the posterior surface of the lens, is the parameter necessary to simulate the movement of the eyes behind the lens and optimizing the visual areas. This vertex distance is of approximately 12 to 14 mm, but it varies from person to person and there are many cases in which the same person has different vertex distances for each eye, separately.

The vertex distance is assumed in standard lens manufacturing of 12 mm and a positioning of the glasses frames closer or further away from your eyes, more or less influences, depending on the patient, the visual comfort (see Fig. 1) [3].

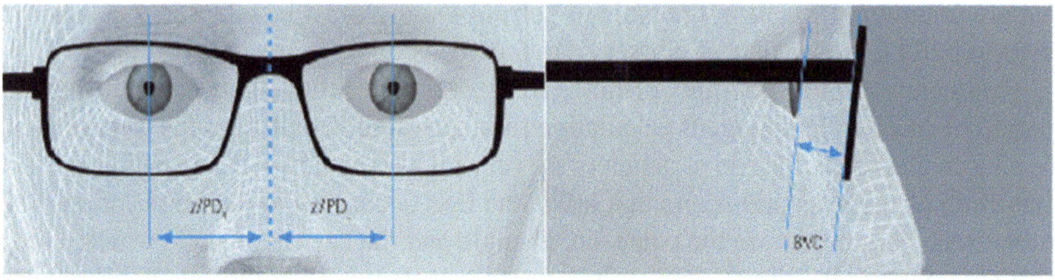

Fig. 1. The pupillary distance on the center right and left interval, measured from the center of the nasal pyramid/Vertex distance - the distance between the plan tangent to cornea and the posterior tangential plan of the lens [3].

1.3 Convergence

Convergence is described by the movement of the eyeball around its center of rotation. This movement of the eyeball occurs when the eye move the sight from infinity to close spaces. When we read or work within a small distance, convergence is maximum, pupils approaching the nose (see Fig. 2). This movement of the eyeball is not identical in all individuals, convergence is more pronounced or not, depending on several factors such as the diameter of the eyeball, power accommodative various pathological conditions, etc. There are people who don't have convergence. This rotation of the eyeball is of about 1.5 mm. Some special devices measure convergence on each eye.

Fig. 2. Convergence – looking to an object from a close distance.

Convergence insufficiency is an imbalance in maintaining the binocular convergence. Paralysis of the convergence or the lack of convergence occurs after cranial-cerebral trauma, encephalitis, stroke, multiple sclerosis, diphtheria. It is characterized by double vision only for the close sight.

2 Frame's Parameters

The frame is very important and it is part of the visual system of the patient. The frame will be chosen according to physiognomy, to the material it is made of and depending on the lens that will be part of the assembly (see Fig. 3) [3]. The frame will be comfortable and must not slide off the face of the patient. For this there are taken into account the cheeks shape, the nose position, the eyebrows, and not least the eyes and ears position because there are fixed the support arms of the glasses frames.

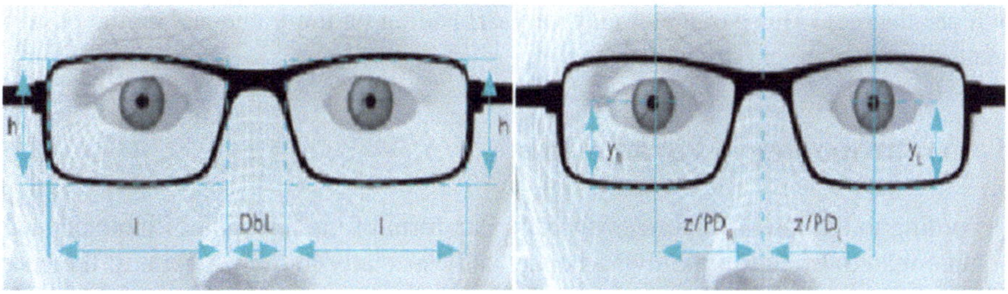

Fig. 3. Constructive parameters of the glasses frame – front view [3].

In addition to the size and shape of the chosen frame, the parameters of the frame offer final details for a balance between the size and optical performance areas individualized visually and aesthetically. The variable height centering allows the selection of eyeglass frames of any size and allows the lens to be designed accordingly. A precise measurement of the inter-pupillary distance secures the visual area for the close sight.

In building of the customized progressive lenses there are taken into account all these design parameters and the following parameters for final positioning of the assembly.

2.1 Curvature Angle of Frame

This angle is described by the curvature radius of the two catches of the frame. For the glasses frames commonly used, this angle has little value between 0–15 degrees. The angle is greater for the frames used in sport or for protection, because it should cover a wider area and follow the shape of the face (see Fig. 4).

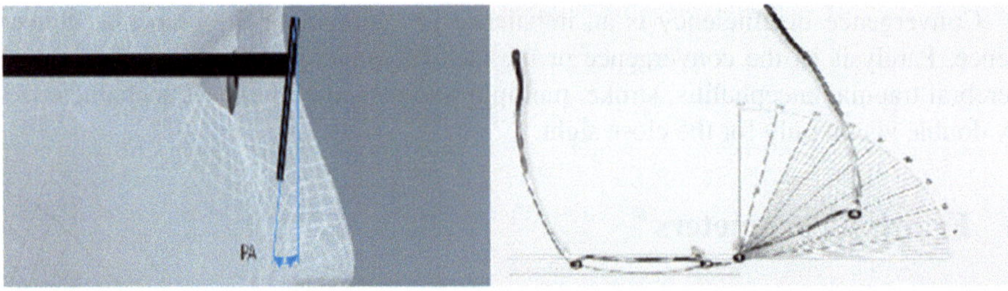

Fig. 4. The pantoscopic angle of the formed optical system/Curvature angle of the frame [3].

2.2 Pantoscopic Angle

The pantoscopic angle is the angle determined by the frame plane to the perpendicular plane to the optical axis of the eye looking to infinity; this angle has the standard value of 8 arc-degrees. This parameter may vary depending on the frame and on the physio-anatomical parameters of the patient.

3 Anthropometric Parameters

According to the laws of nature, symmetry is a form of organization of inorganic and organic materials. But in the living beings and especially in the vertebrates, the body symmetry is not a constant.

In particular it was found that body shapes and especially human facial shapes are usually uneven - this situation depends on the intrauterine position, vasculature and neurotroficity stimulation during the development period, hemispheric dominance that favors the skeleton and muscles of the dominant hemi-body.

3.1 Standard Facial Parameters

Although the slight facial asymmetry represent natural situations, they are usually hard perceptible. Even champions or world champions have small inequalities in terms of beauty of face shapes: uneven nostrils, small differences in the position of the ears, insert appendages, eye, etc.

The proportions between these characteristic dimensions, together with the color and facial figure, determine the individual physiognomy. To accurately restore these proportions, there are used several anthropometric landmarks consisting in certain points, lines and planes located between the face and the figure limits (see Fig. 5).

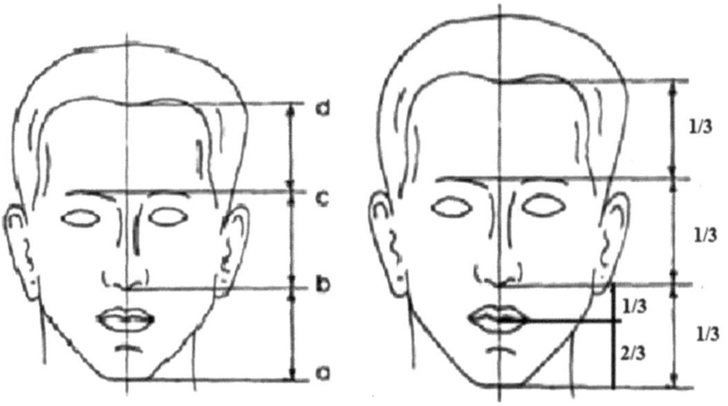

Fig. 5. Delimitation of the face and figure/standard aspect ratio of the face and figure [3]. a-b-c = face; a-b-c-d = figure; a-b = inferior side of the face and of the figure; b-c = superior side of the face and middle side of the figure; c-d = superior side of the figure.

3.2 Facial Asymmetry Raises Issues in Construction of Glasses Lenses

The facial asymmetries have neurological implications, genetic and neuro-vascular.

These syndromes are to be found both in adults and in childhood, puberty and adolescence. The neurologist knows the cranial nerve function and the autonomic nervous system, so he is the first who can determine the difference between anatomical and pathological variants. Anthropometric notes on facial abnormalities begin with standard measurements of the head and facial proportions, existing broad theories on normal and extreme faces to be framed in dysmorphia (see Fig. 6) [4].

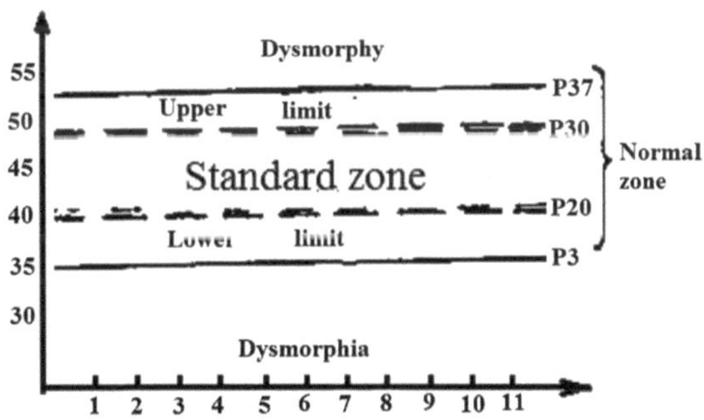

Fig. 6. Scheme of the normal territory and facial dysmorphia according to the anthropometric measures. S. Stengel Rutkowski – 1985 [4].

In a classification commonly found in the medical literature, we find more than 100 craniofacial abnormalities marked by the morpho-anthropologic appearance, and among these there are described 25–28 with specific facial dysmorphia. Genetic facial abnormalities and embryonic malformations are clinically detected as facial cracks, atrophies or facial abnormalities or defects of the first pharyngeal mesoderm arch. Agenesis and dysgenesis of the branchial arches can affect the facial massif in whole or only on a half face. All these situations are facial asymmetries and defects [4].

Septum deviations, the position in different planes of the eyeballs both vertically and horizontally, deformation of the cheeks, different positioning of the ears, are problematic in wearing a pair of glasses that leads to a special construction of the lens that takes into account all patient parameters.

4 Measurement Techniques of Anthropometric and Final Spectacles Parameters

In time, it was wanted that the anthropometric parameter measurements of the patient to be as exact as possible. This means a very precise construction of glasses and of course maximum comfort for the wearer. It started out as a distant interpupillary ruler, then the measurements were made mechanically and digitally. A more precise ruler is the one that measures pupil position in relation to the vertical plane of nose symmetry in the area of the frame of the eyeglass. The user place the special cutting on the middle of ruler on the nose base and measure from the zero point to the center of the left pupil and then to the center of the right pupil.

Such measurements are necessary due to multiple measurement errors with simple ruler - from pupil reading errors due to low contrast between pupil and iris, eye movements, oblique reading and parallax error (removing it requires an eye to be covered by the patient and optometrist and open eye-to-open eye reading, which slows down the process and may cause problems for less well-meaning patients, the elderly or the very young) to issues of poor communication between the patient and the optometrist, or to the fact that it is difficult to control the direction of the patient's eye and the distance to the target (see Fig. 7) [5].

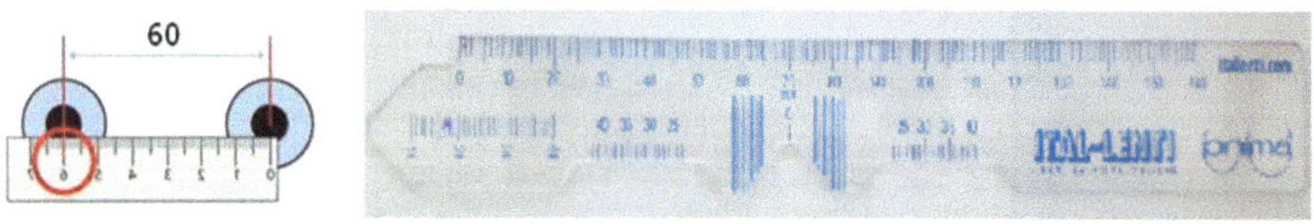

Fig. 7. Measurement ruler for the pupillary distance/special ruler for measuring pupil distance on left and right intervals [5].

With pupilometer is possible to solve the problem of distance to the target and the measurements are made for each eye, but it remains the problem of patient cooperation (it has fewer opportunities to distort the measurement but still exists). It is necessary to directly contact the device with the face of the patient but, most importantly, the measurements are made in a forced state (by aligning the sight on a target in the device and having the right head), which may differ from its natural state of sight. At the same time, it is not possible to measure the positions in relation to the frame, which is very important both for the reduction of the errors in the calculation of the lenses but also because the conditions of the look change as the transition from the unobstructed look to the look in the compensating conditions and the glass eye.

Technology has advanced rapidly so precise devices have been built that can measure all the necessary parameters (see Fig. 8) for personalized progressive lenses by taking photos.

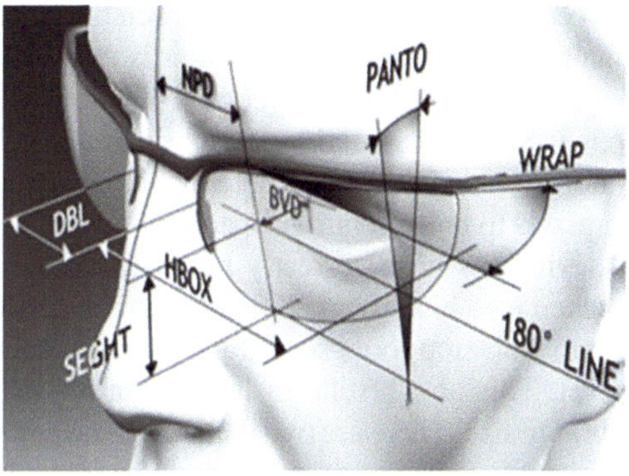

Fig. 8. Individual parameters of the patient, inclusive of the selected frame [3].

All these individual parameters are used in the construction of the progressive lens in order to personalize it: interpupillary distance in the range, the vertex distance, pantoscopic angle, the convergence of each eye, radius of curvature of the frame, the height of installation, nasal bridge.

Measurements are made with a device that comprises specially designed software, processing images of patient face with spectacles (see Fig. 9).

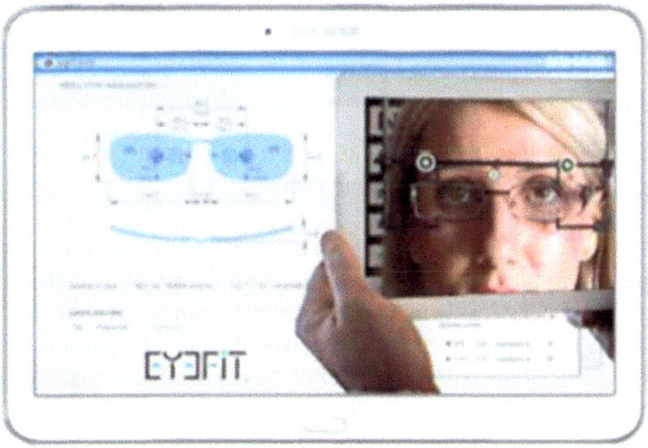

Fig. 9. EYE FIT mobile device to determine the individual parameters [6].

Very important for making measurements is the frame clamp device, which has three markers positioned so that all measurements are calculated in relation to their layout.

The device is made of carbon and has 9 grams, making it easy not to change the position of the eyeglass frame it is catching. All dimensional and constructional and guiding angles of the device are known and will be part of the calculation of the measurements to be run by the software. It is very important to grip and position the device with the three guides, over the frame of the glasses chosen by the patient. Positioning is done taking into account that the marker in the center is exactly half the face and the selected frame. The device is required to build a correct position of the coordinate system in which the measurements are made and to establish image correction parameters. Without markers it is not possible to correctly determine the coefficients of the passage matrices from units in pixels to units in mm in space.

Once the marker device has been correctly mounted on the eyeglass frame, the patient's pictures are taken (see Fig. 10). The first picture is performed in front of the patient with his eyes in the tabletop room so that he can calculate the pupil distance to infinity for each interval. This also results in lens mounting height, eyeglass frame dimensions and nasal bridge size. There are two pictures in the profile, with the patient's eyes towards infinity, pictures that determine the frame pantoscopic angle, the frame curvature and the vertex distance for each eye. The last picture will be taken with the patient in close proximity to the tablet room to measure the following parameters: convergence on each eye and lens read position.

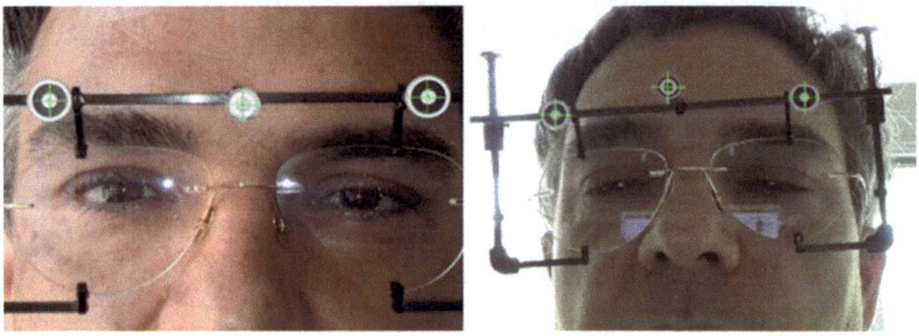

Fig. 10. Example of images captured during measurement with EYE FYT - individual parameters – far sight and close sight measurements

The software that runs on such a device is constructed to recognize numerous pixels forming the picture to turn the unit of measurement in mm. The precision of measurements made with these devices is in terms of tenths of mm. After the patient was photographed, the EYE FIT program runs the received data and as a result we have all the patient measurements of the parameters and of the chosen frame parameters in order to build an ultra-personalized progressive lens (Figs. 11 and 12).

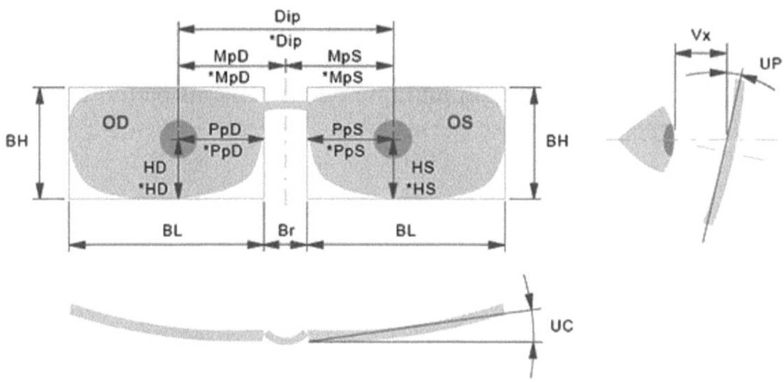

Fig. 11. Example of close-sight individual parameters with EYEFYT after processing images

						*Close sight distancees		
Dip:	70.7	mm	BH:	42.5	mm	*Dip:	*68.3	mm
MpD:	34.9	mm	BL:	55.1	mm	*MpD:	*33.9	mm
MpS:	35.8	mm	Br:	17.7	mm	*MpS:	*34.4	mm
PpD:	25.7	mm	Vx:	22.15	mm	*PpD:	*23.1	mm
PpS:	26.8	mm	VxOD:	22.7	mm	*PpS:	*23.5	mm
HD:	30.8	mm	VxOS:	21.6	mm	*HD:	*24.2	mm
HS:	28.4	mm	UP:	5.4	°	*HS:	*22.4	mm
			UC:	4.6	°	Reading distance:	*36.5	cm

Fig. 12. Example of measurements of the individual parameters measured by EYE FIT. Reading distance; Dip = interpupillary distance; MpD = interpupillary distance for the right eye; MpS = interpupillary distance for the left eye; Vx = vertex distance (from the cornea to the lens); HD, HS = The distance from the pupil to the tangent in the lens; BH = Height of the frame; BL = Frame length; Br = bridge; UP = pantoscopic angle; UC = frame angle of curvature

Differences of parameter measurements are observed between the two halves of the face. For people with large differences in parameters, special lenses are built to give them comfort and visual field to meet their work needs.

5 Conclusions

Technology is constantly evolving. The help that man receives from advanced technological equipment is substantial. In the field of ophthalmology, technology helps us in accurate determinations of pathologies not detected with the naked eye. Technological evolution shows that new knowledge and technological advances in image acquisition, computing power in portable devices, the explosive development of mathematical fundamentals and image processing algorithms, and non-contact spatial measurements have allowed the passage of to manual measurements affected by numerous errors, to automatic, performance measurements.

However, it should be noted that there are still major issues related to patient collaboration, uneven illumination, the need to achieve several successive pictures with patient resting without any movement and their processing.

Consequently, advanced studies are needed to improve the method, to reduce the number of shots (by improving algorithms to be more robust in relation to error sources and to making an optical system with multiple micro-cameras permitting the simultaneous execution of more photos).

Acknowledgement. The authors thank Nova Lenti company and, in particular, Mr. Mihai Leu for permanent support during this study.

References

1. Dumitrescu, N.: Bazele opticii fiziologice, Ed. Politehnica, Bucureşti (1991)
2. Bioptical art Homepage: http://www.artabioptica.ro/art3.htm. Accessed 21 May 2018
3. Technical-commercial documentation/catalogue - Carl Zeiss
4. Stengel-Rutkowski, S., Schimanek, P.: Chromosomal und nicht-chromosomale Dysmorphie syndrome. Enke Verlag, Stuttgardt (1985)
5. Ophtalmi'x Blog Homepage: https://oftalmix.wordpress.com. Accessed 21 May 2018
6. Soft EYE FIT - © CENTWINS SRL: www.optivio.ro. Accessed 21 May 2018

Study on the Current State of Research in the Field of Artificial Clamping Systems Using Astringent Electroadhesion Technology

Mihai-Nicolae Popescu and Mihai Avram(✉)

Politehnica University of Bucharest, Bucharest, Romania
mihai.p.nicolae@gmail.com, mavram02@yahoo.com

Abstract. This paper aims to provide an analysis of the current state of research in the field of artificial clamping systems using astringent electroadhesive technology.

The principle of electroadhesion phenomenon is described based on devices build as research prototypes made by several technology developers and research institutes.

Here is presented, starting with the first uses of electroadhesion devices in the 1960s as a standalone technology by NASA Research Center on astronaut garment for outer space, to the prehensile prototypes developed by Laboratory of Intelligent Systems, School of Engineering, Ecole Polytechnique Federale de Lausanne, used for gripping soft, fragile objects.

In this paper we show the various advantages of electroadhesion devices, also the limitations that depend on the proprieties of the adherent object.

Focusing on the reliability, the materials and the operating environment in the future commercial users(factories, logistics centers etc.) could envision the replacement of classic devices based on vacuum or magnetic technology with electroadhesive devices.

Keywords: Electroadhesion · Clamping system · Electroadhesive

1 Introduction

The electrification of bodies has been known since antiquity.

The Greek philosopher Thales of Miletus, who lived 2500 years ago (7th century BC), is the first to recall the phenomenon in his writings, where he presents experiments with pieces of amber which, when rubbed with a piece of fur, cause a electrical charge on its surface, attracting light objects [1].

The first commercial use of electroadhesion was at the paper-sheet fastening system of HP 9125A plotter (see Fig. 1).

G. I. Gheorghe (Ed.): ICOMECYME 2018, LNNS 48, pp. 149–155, 2019.
https://doi.org/10.1007/978-3-319-96358-7_15

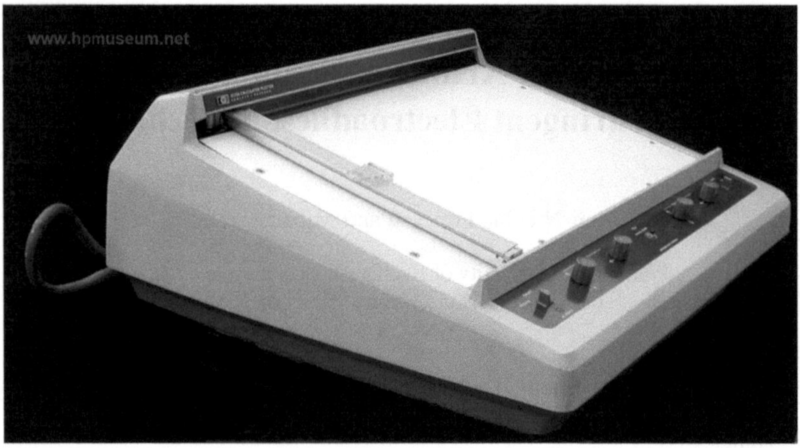

Fig. 1. HP 9125A plotter

With the advancement of technology in the electronics industry, different uses of this phenomenon have appeared in the handling of fragile and scratch-sensitive objects.

In the field of artificial clamping systems with astringent action on objects, there are several types of technologies used:

- vacuum
- magnetic
- electroadhesive

"Electroadhesion is the electrostatic effect with astringent action between two surfaces subjected to an electric field" [2].

Electroadhesive phenomenon is expressed by the formula:

$$\frac{F}{A} = \frac{\epsilon V^2}{2d^2}$$

where:

- F is the strength between two charged electrodes separated by an insulator in the electroadhesive bond;
- ϵ is the dielectric constant of the insulator;
- d is the measurement of electrode separation distance;
- A is the contact area [3].

The principle of operation of electrically-adhesive devices varies according to the conductivity of the object over which prehension is exerted (properties represented by the simplified expression of Gauss's law):

(a) for objects from dielectric materials $D = (\epsilon E + P)$
(b) for objects from conductive materials $D = \epsilon E$

The capability of supplying electric power to the electrodes by a circuit gives to the electroadhesion devices the advantage of electrically controllable adhesion technology by a single control voltage.

The constructive way encountered in experimental works is represented by conductive electrodes isolated in a polymer (Fig. 2).

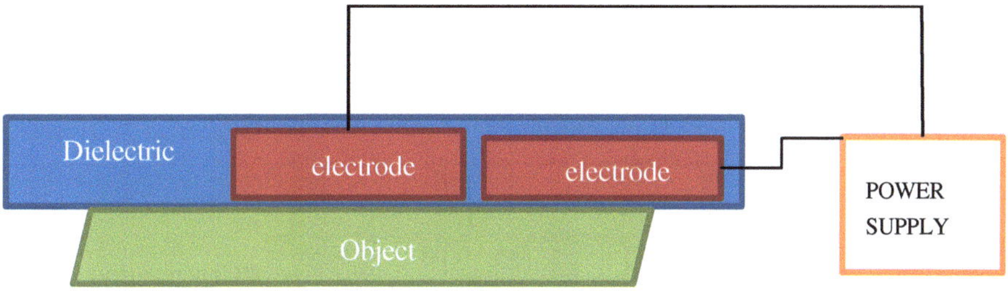

Fig. 2. Graphical representation of electroadhesive pads

The electric fields set up opposite charges on the substrate when an electric charge is applied and thus the electrostatic adhesion between the electrodes and the induced loads on the substrate occurs.

2 Applications of Electroadhesive Phenomenon

2.1 Tracked/Wheeled Robots

SRI international has developed and envisioned several types of wall climbing robots, one of them registered as patent US20100059298A1 (Fig. 3).

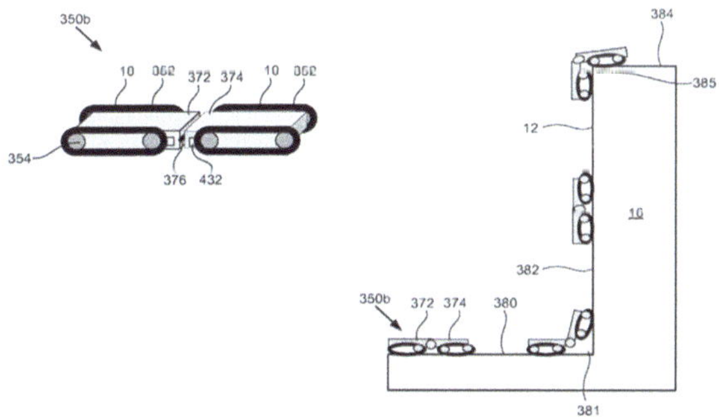

Fig. 3. Graphical representation of the wall-crawling robot [5]

Described as a robot that uses for locomotion a mechanism that also comprises of an electroadhesive device, thus creating an adhesive force capable of obtaining vertical mobility.

Based on the clamping performance in testing of the robot on different materials, they have measured the pressure obtained by the electroadhesive pads at a 4 kV DC voltage with the following results (Fig. 4).

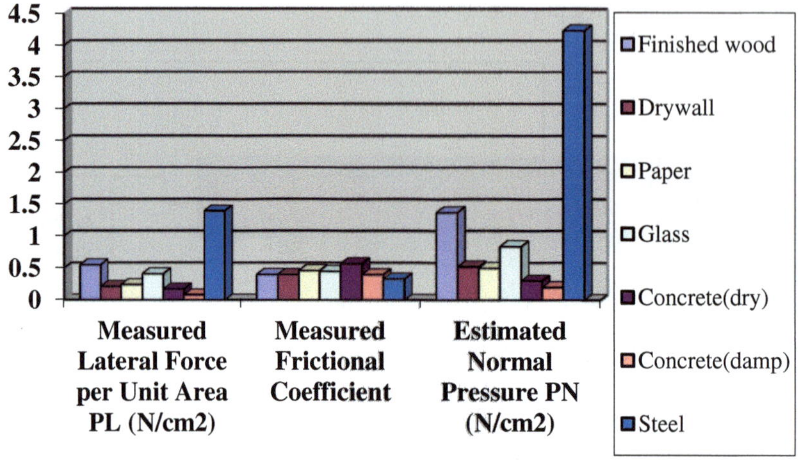

Fig. 4. Clamping performance obtained on different materials [6]

2.2 Prehensile Devices

2.2.1 Garment Application

In the paper "Application Study of Electroadhesive Devices" by Richard P. Krape prepared by Chrisler Corporation Space Division for National Aeronautics and Space Administration Langley Research Center [3], electroadhesion devices are started being researched and developed for their potential usefulness in prehension applications.

They used the electroadhesivity phenomenon in developing prototypes with multiple uses in aerospace garment applications as hand holds, knee pads, back and several other space suit accessories.

Specifications of the prototype (Fig. 5), at 1850 V, the static pull force is 2.7 kgf and the skid force is 18.14 kgf with a battery longevity for operation of 42 h.

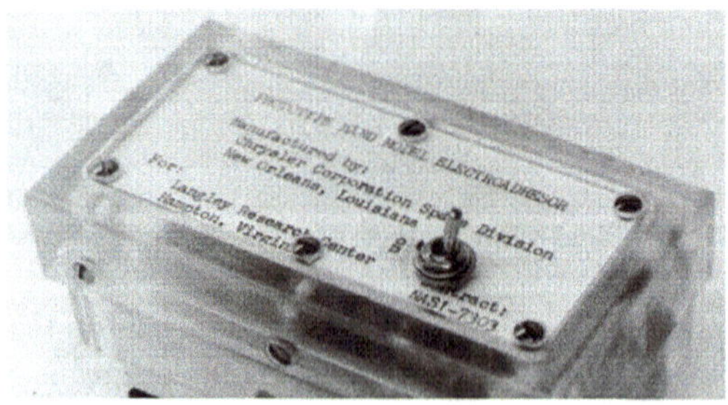

Fig. 5. Hand-Model Electroadhesor Prototype – Single Pole Type [3]

Specifications of the prototype (Fig. 6), at 1850 V, the static pull force is 0.9 kgf and the skid force is 0.4 kgf with a battery longevity for operation of 40 h.

Fig. 6. Flexible Prototype (Single-Pole) [3]

Following their research, they determined that the resistivity, the chemical nature, the temperature, polarity of applied voltage, current density and the thickness of the electroadhesive coating influence the proprieties of electroadhesion devices directly.

2.2.2 Stretchable Electroadhesive Pads

Based on these proprieties electroadhesive devices have been developed in the form of stretchable electroadhesive pads. In this case, the forces that generate electroadhesion vary depending on how the pads are deformed [7] (Fig. 7).

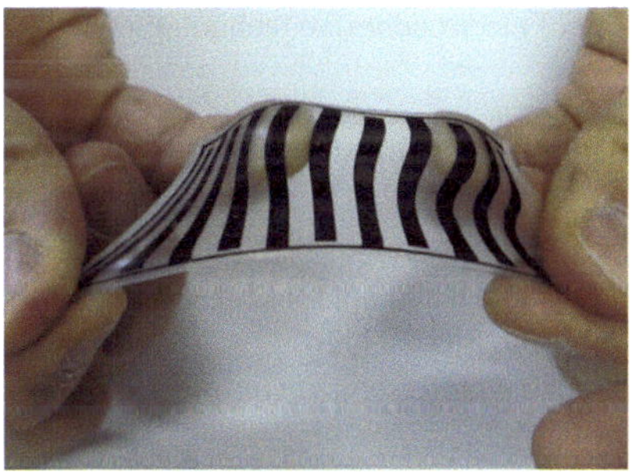

Fig. 7. Stratchable electroadhesive device [7]

If the electroadhesive pads are stretched, the space between electrodes increases, decreasing the electric field strength thus resulting in decreased polarization.

Stretchable electroadhesive devices are built in 2 stages using a rubber mixture having molding capability:

- first stage: the conductive rubber mixture containing Carbon black is poured into the mold obtaining the electrodes
- second stage: the electrodes are encapsulated using rubber poured over them as insulator.

2.2.3 Soft Grippers

Using stretchable electroadhesive devices technology there has been developed a new method for creating soft grippers, enabled to operate deformable and fragile objects of any shape.

According to the creators of (Fig. 8), the gripping force produced by electrostatic actuation is low (1 mN), allowing operation of fragile objects, while the electroadhesion force can produce around 3.5 N/cm^2, thus enabling the capability of heavy objects lifting.

Fig. 8. Soft grippers [8]

3 Conclusion

The purpose of this paper is to present the status of research in the domain of electroadhesion clamping devices, a short history on how this technology has been put to use, the advantages and the limitations of this technology as shown in Table 1.

Table 1. Comparison of electroadhesive technology with vacuum and magnetic

Technology	Advantages	Limitations
Vacuum	High adhesion forces on smooth surfaces	Noisy, energy inefficient on rough surfaces; requires separate pumping source
Magnetic	Low-energy cost when not moving; quiet operation; Proven high adhesion forces on ferromagnetic materials	Technology available only for ferromagnetic surfaces; high-energy cost when moving; cannot climb some plastic surfaces
Electroadhesion	Adheres to a large variety of surfaces; has low-energy cost when mobile; it is quiet, simple, lightweight, compliant; it can be switched off for cleaning	Requires very small amounts of power (~0.02 mW/N of weight supported) to stay clamped

Worldwide it is estimated that in the future most jobs will be taken over by robots, but some of the jobs they will replace are hazardous jobs, by using them we will improve human wellbeing, this vision of future is very well described by Larry Page, Google general manager when he stated that "robotics will improve work and allow people to spend more time with their family" [9].

This is going to become a starting point for the authors in developing new robotic applications using the electroadhesive phenomenon.

References

1. Trevena, D.H.: Static Fields in Electricity and Magnetism. Butterworths, London (1961)
2. Monkman, G.J.: Workpiece Retention during Machine Processing. Assem. Autom. **20**(4) (2000) (MCB University Press)
3. Krape, R.P.: Applications study of electroadhesive devices. Chrisler Corporation Space Divison for National Aeronautics and Space Administration Langley Research Center (1968)
4. Gaunekar, G.S.: Implementation of electroadhesion based astrictive prehension system. Int. J. Adv. Res. Electron. Commun. Eng. (IJARECE) **5**(11) (2016)
5. Pelrine, R.E., Prahlad, H., Kornbluh, R.D., Lincoln, P.D.: Scott Stanford, SRI International. US Application, US20100059298A1
6. Prahlad, H., Pelrine, R., Stanford, S., Marlow, J., Kornbluh, R.: Electroadhesive robots—wall climbing robots enabled by a novel, robust, and electrically controllable adhesion technology. In: 2008 IEEE International Conference on Robotics and Automation, Pasadena, CA, pp. 3028–3033 (2008)
7. Germann, J., Schubert, B., Floreano, D.: Stretchable electroadhesion for soft robots. In: 2014 IEEE/RSJ International Conference on Intelligent Robots and Systems, Chicago, IL, pp. 3933–3938 (2014)
8. Shintake, J., Cacucciolo, V., Floreano, D., Shea, H.: Soft robotic grippers. Adv. Mater. 1707035 (2018). https://doi.org/10.1002/adma.201707035
9. https://www.theguardian.com/technology/2014/jul/07/google-founders-larry-page-sergey-brin-interview

Modelling the Tibial Bone Using CAD Techniques, Starting from the 3D Scan Model

Mihai-Constantin Balaşa[1,2], Simona Mihai[3(✉)], Viviana Filip[2], Alexis-Daniel Negrea[4], and Gheorghiţa Tomescu[5]

[1] Macartney Hydraulics A/S, 7620 Lemvig, Denmark
[2] Doctoral School of Engineering Sciences, Valahia University of Targoviste, 130105 Targoviste, Romania
[3] Institute of Multidisciplinary Research for Science and Technology, Valahia University of Targoviste, 130004 Targoviste, Romania
mihai.simona@yahoo.com
[4] Faculty of Material Engineering and Mechanics, Valahia University of Targoviste, 130004 Targoviste, Romania
[5] Faculty of Mechanical Engineering and Mechatronics, University Politehnica of Bucharest, 060042 Bucharest, Romania

Abstract. To improve the long-term quality of life for patients with major joint problems, choosing an orthopedic implant based on patient particularities is essential. In this sense, numerical modeling of the bone and implant is useful, assembling them and simulating the behavior of this ensemble over time, taking into account the most accurate features of the bone, such as varus - valgus deformities, osteoporotic diseases, etc. The need to use CAD 3D modeling techniques thus appeared. Because bone is irregularly shaped, consisting of more complex 3D surfaces it is difficult to build in SolidWorks. That's why, to build it, we started from the virtual 3D scan model with the ExaScan scanner. Once the virtual model has been obtained through CAD techniques, it will be adapted to the patient's particularities (dimensions, bone mass, eventual deformities, etc.).

The objective of this paper is numerical modeling of the tibial bone using CAD techniques, starting from the virtual model obtained by 3D scanning.

Keywords: Tibial bone · Knee implant · Virtual model · 3D scanning CAD techniques

1 Introduction

Knee prosthesis is the surgical procedure by which the affected knee cartilage and the bones are removed and replaced with an implant which allows for natural movements of the knee joint (primary surgery) [1, 2]. Although implants have a relatively long life, but limited, sometimes problems occur after a relatively short time and an intervention should be performed to replace the original prosthesis with another prosthesis [3]. This surgery, called revision surgery, is more difficult than the primary surgery because the prosthesis is fixed in the bone tissue, and the surgeon, by removing the initial implant, will also remove some of the bone in which it is fixed [4]. Therefore, it would be of great

G. I. Gheorghe (Ed.): ICOMECYME 2018, LNNS 48, pp. 156–165, 2019.
https://doi.org/10.1007/978-3-319-96358-7_16

use if before performing the primary surgery, the surgeon would know how the implanted bone behave [5], depending on the type of implant, the particularities and affections of the patient [6], in order to personalize the best possible the implant for the patient and to avoid or postpone the revision surgery as much as possible [7, 8].

The aim of our work is to achieve the virtual model of the tibial bone, which we then use for finite element analysis of the knee implant bone assembly in order to determine its behavior over time.

2 Working Method and Resources Used

We created the virtual model of a tibial bone 320 mm long, with a cortical bone thickness of 0.8 mm (at the extremities toward the knees and toward the ankle) and 4 mm (in the middle).

First, we performed the scanning of the real tibial bone (Fig. 1) with an ExaScan 3D scanner [9] and obtained a complex 3D surface, Fig. 2, which we converted into a solid body. The result of scanning has a very good fidelity due to the large number of surfaces/meshes obtained [9]. But this very multitude of surfaces has created difficulties in the finite element analysis of the tibial-knee implant assembly.

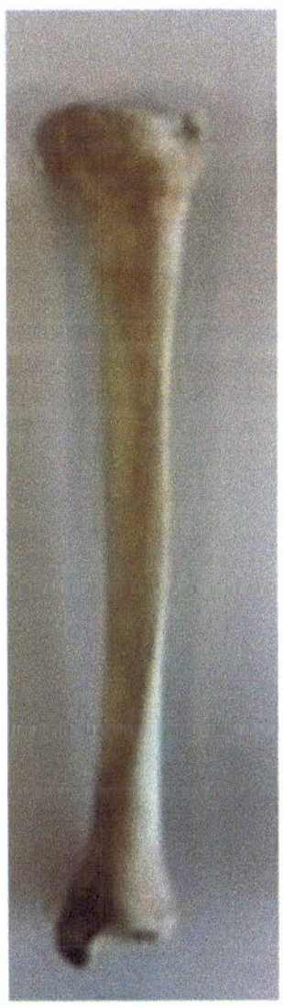

Fig. 1. Tibia, real bone.

This is why we decided to achieve the virtual model of the tibial bone using CAD modeling techniques in SolidWorks [10, 11]. But this is very difficult due to the complexity of the shape of the tibial bone (many irregular surfaces with complex profile). In order to succeed, we used the surface obtained by 3D scanning, processed in CATIA (Fig. 2b) [9], which we named reference model.

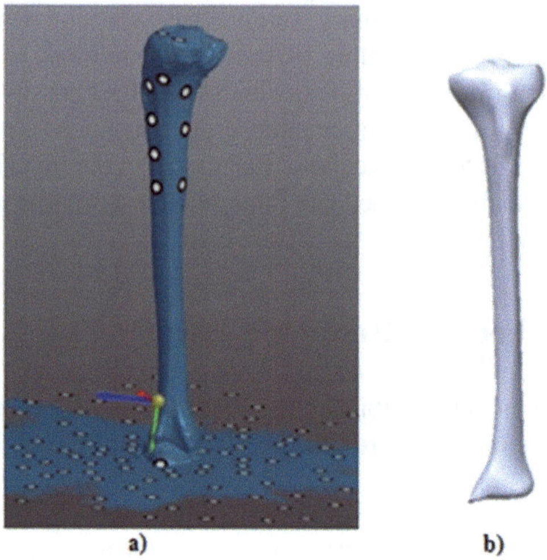

Fig. 2. The 3D surface of the tibial bone, obtained by scanning (a) raw scanning, (b) processed scan in CATIA.

The use of the reference model greatly facilitates the shaping of the tibial bone. In the following, we will use the model in work term for the virtual model that we want to achieve.

The steps are:

Step 1

We established the tibial bone axes for both the reference model and the model in work, to work in correspondence (Fig. 3).

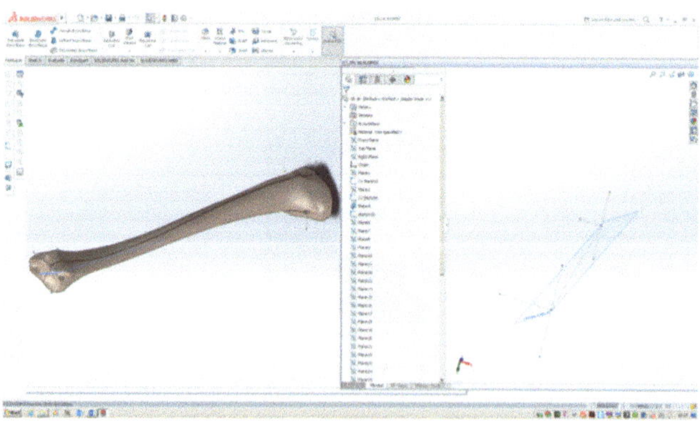

Fig. 3. Choosing plans and work axes (reference model on the left, model in work on the right).

Step 2

We cut the reference model with parallel planes with a distance of 10 mm inbetween (transverse slices). A contour resulted in each section plane. We created a sketch in each plane, in which using a Spline command, Fig. 4, we drew a curved overlay on the contour previously obtained This built curve was extracted and taken with the Copy command in the corresponding plane of the model in work and the action was repeated for each section plane (Fig. 4).

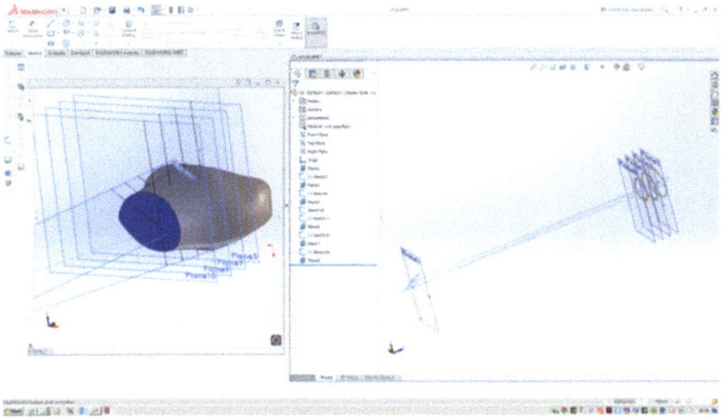

Fig. 4. The processing of the tibia contour in successive plans from the reference model (left side) and it's implementation in the model in work (right side).

For a better accuracy when taking these curves which define the bone profile at the proximal end of the tibia, the plans with which we cut the reference model were created at a smaller distance, namely 5 mm (Fig. 5). We mention that we performed the slicing of the reference model without the end of the tibial bone, as they are not necessary for the bone knee implant assembly.

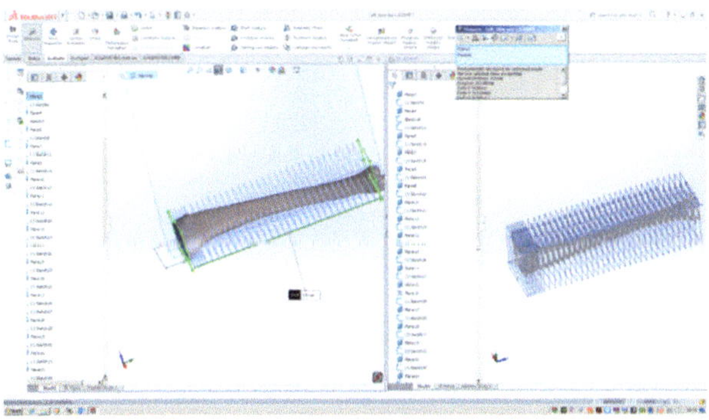

Fig. 5. Slicing with parallel planes (reference model on the left, model in work on the right).

Step 3

Slicing with parallel planes (reference model on the left, model in work on the right) the Loft command with which we will build the bone in the model in work (Fig. 6).

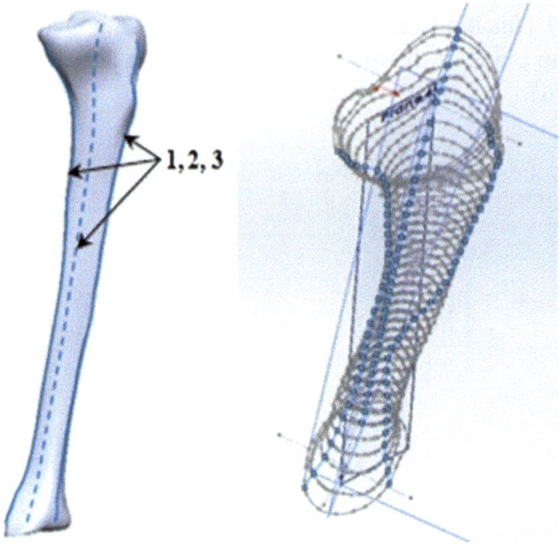

Fig. 6. The three representative curves defining the tibial surface (reference model on the left, model in work on the right).

In the reference model we opened a sketch for each of the three curves and using Spline we drew, overlapping guiding lines over the 3 curves. Using the Copy command, each guiding line was transferred to the model in work, and then, with the Pierce command, we made the guidance line intersect each contour curve in the parallel planes previously described (Fig. 7). We repeated the process for each guidance line.

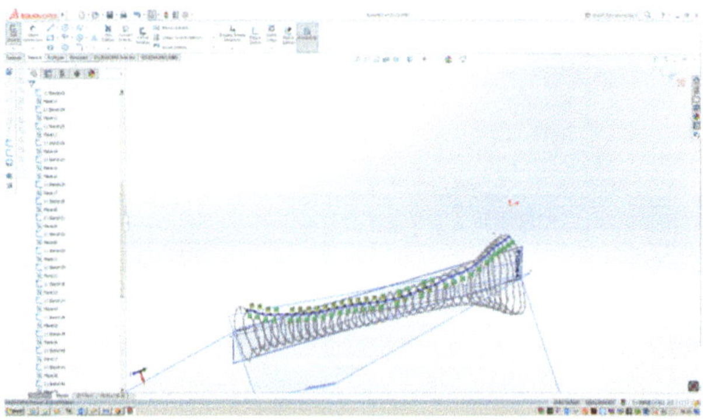

Fig. 7. Modeling guide lines and fixing them on contour curves

Step 4
We created the 3D modeling of the bone using the Loft command by selecting all the contour curves and the three guiding lines (Fig. 8).

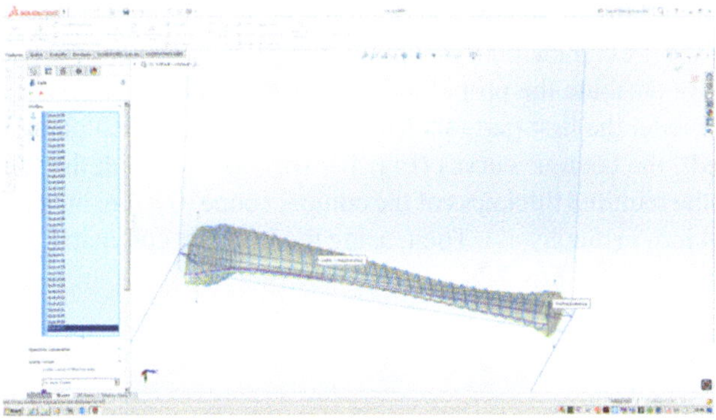

Fig. 8. Obtaining the 3D model

To this point we achieved the tibial bone as a solid body (full). In order to perform the finite element analysis, it is necessary to assign distinct material properties to the bone [12], depending on the bone area, taking into account the following aspects (Fig. 9): long bones (such as the tibia) are made up of a compact bone shell, they have a medullary canal in the center, and at the two extremities (proximal epiphysis and distal epiphysis) they have a block of spongy substance, surrounded by a layer of the compact substance [13, 14].

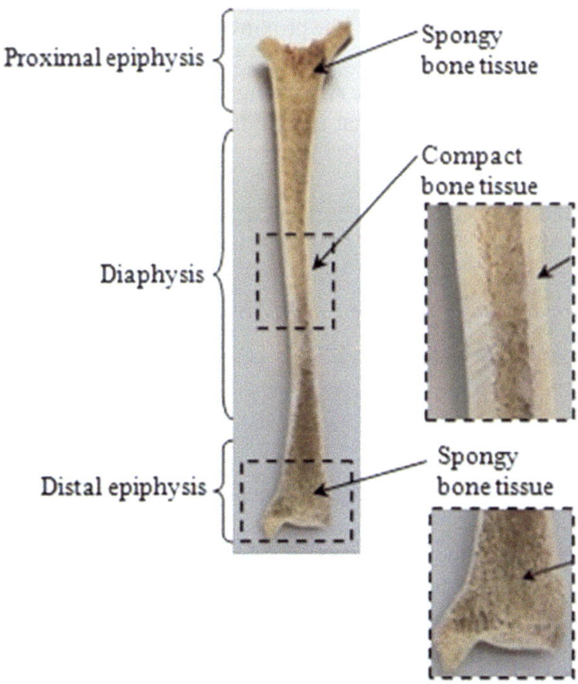

Fig. 9. Tibial bone structure

For this, we will divide the obtained 3D model in three parts (Fig. 9): the shell (to which we attribute the compact/cortical bone properties), the proximal part and the distal part (to which we attribute the properties of the spongy bone).

In order to model the first part (shell), in each sketch (in each plane with which the bone was sliced), the contour curves (Fig. 10) was doubled with the Offset command, thus obtaining the required thickness of the compact bone (0.8 mm in proximal and distal epiphysis and 4 mm in diaphysis). Then, using the Loft Cut command, the virtual model of the shell was obtained (Fig. 12a).

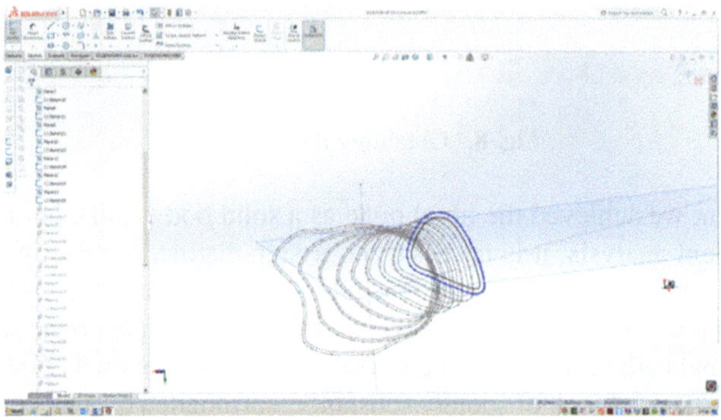

Fig. 10. The doubling of the contour curves with the purpose of building the shell

To model the second and third part (the proximal and distal side), we use the sketches built for the outer layer and to build offset guides parallel to those used in the shell construction with the Offset command. Also using the contour curves and with the Pierce command, we make the constraint that the guiding lines intersect each inner contour curve (Fig. 11).

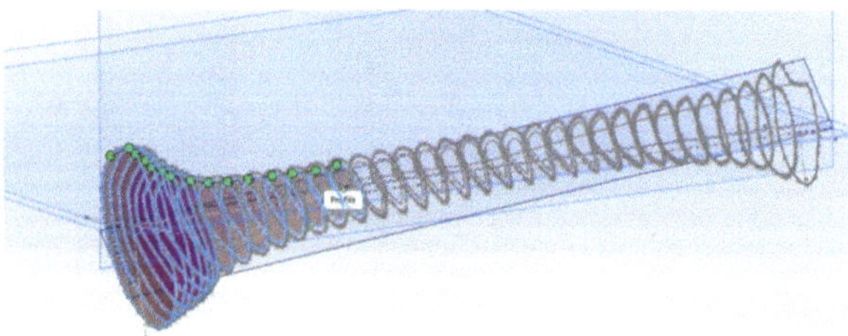

Fig. 11. Fixing the guiding lines on contour curves

Using the Loft command, we generate a solid that has the shape of the tibial bone but with a lower thickness (the coating has been removed). The 3D model obtained is saved twice, thus obtaining two identical models. In the first model, at a distance of 100 mm from the end of the knee, we drew a sectional plan and used the Cut-Extrude command to retain only the proximal part, Fig. 12b. In the second model, at a distance of 100 mm from the end of the ankle, we draw a cutting plane and used the command Cut-Extrude to retain only the distal Fig. 12c.

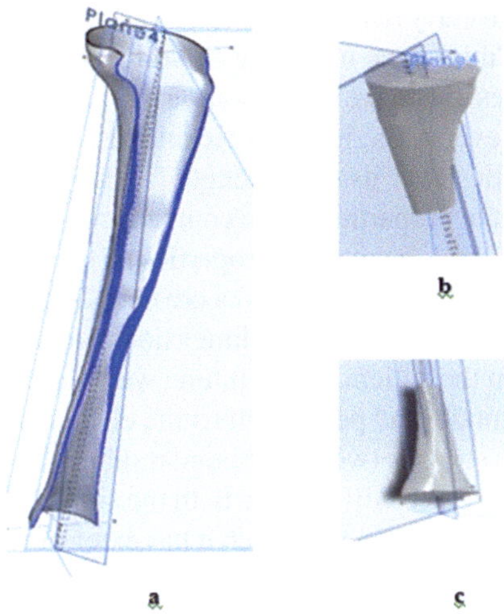

Fig. 12. The shell (a), the proximal part (b), the distal part (c)

In order to obtain the virtual model of the tibial bone, we assemble the proximal and the distal part in this shell, Fig. 13.

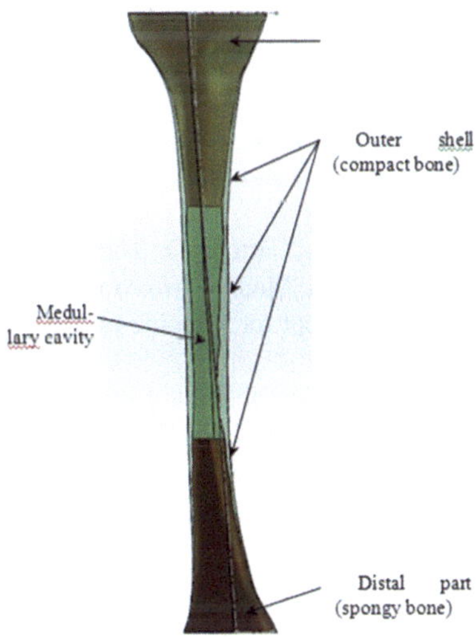

Fig. 13. Assembling of the three parts of the virtual model

In this way, we can assign materials with distinct properties to the three parts.

3 Conclusions. Future Research Directions

In this paper we have developed the virtual model of the tibial bone using CAD techniques, which was for us to carry out the finite element analysis of the tibial knee implant assembly in the future. Tibial bone modeling in SolidWorks was not an easy task due to irregular shaped surfaces that make up the bone. That's why we started from the virtual 3D model [9], through Reverse Engineering techniques. Since the bone does not have the same material properties over its entire length [5], we have created three parts: the shell (which will receive the properties of the compact/cortical bone), the proximal part and the distal part (which will receive the properties of the spongy bone), and through the assembly the virtual model of the tibia was created. Once the virtual model is built in SolidWorks, it will be customizable (dimensions, material properties, eventual deformities) depending on the patient. In the future, we plan to place the cemented tibial plate on the tibial virtual model and perform the finite element analysis of this ensemble, which we subject to static stresses (weight of human subject in orthostatic position) and dynamic (walking, running). We will analyze both the situation in which the tibial plate is without extension, and the situation in which it has extensions of different lengths and thicknesses [15], which are standardized in the profile catalogs.

We will also analyze the situation where the tibial bone has deformations of the varus-valgus type (in which case we apply rotational moments to the finite element analysis) and the situation in which the bone is affected by different degrees of osteoporosis (in which case modify bone density). This numerical simulation of bone-implant behavior is useful for the optimal choice of the shape and size of the implant [16–18], depending on the patient's particularities.

References

1. Cofaru, I.I.: Summary of the Ph.D., thesis, Researches regarding the biomechanics of the axial deviations of the human lower member and the development of the correspondent surgical devices (2013)
2. Cristea, Ş., Prundeanu, A., Groseanu, F., Gârtone, D.: The role of arthroscopy in mini-invasive treatment of tibial plateau fractures. In: Modern Arthroscopy, pp. 225–236 (2011)
3. Mihai, S., Filip, V.: New design concept for reducing torque wear on implant. Appl. Mech. Mater. **658**, 453–458 (2014)
4. Mihai, S., Filip, V., Vladescu, M.: Contributions to the improvement of the tribological behaviour of hip implant joints. J. Sci. Arts **2**(35), 177–184 (2016). (ISSN 1844-9581)
5. Nicolescu, C.M., Bumbac, M., Mihai, S., Gheboianu, A.I., Balasa, M.C., Filip, V., Cuculici, S., Cristea, S., Pantu, C.: X-RAY diffraction and nanoindentation characterization of bone tissue affected by severe osteoarthritis. J. Sci. Arts **1**(42), 265–274 (2018). (ISSN 1844–9581)
6. Mihai, S., Filip, V.: Studies on explanted orthopedic implants. Sci. Bull. VALAHIA Univ. Mater. Mech. **8**(11), 149–152 (2013). (ISSN 1844-1076)
7. Mihai, S.: Constructive concepts for reducing wear caused by the friction torque of hip implants. Int. J. Res. Eng. Technol. **2**, 289–295 (2013)
8. Mihai, S., Filip, V.: Simulator for testing the friction torque of the femoral head of a hip prosthesis. Bull. Transilvania Univ. Braşov **8**(57), 145–150 (2015) (No. 2, Series I: Engineering Sciences ISSN 2065-2119)

9. Balaşa, M.C., Cuculici, Ş., Panţu, C., Mihai, S., Negrea, A.D., Zdrafcu, M.O., Leţ, D.D., Filip, V., Cristea, Ş.: Using 3D scanning techniques in orthopedic systems modeling. Sci. Bull. VALAHIA Univ. Mater. Mech. **15**(13), 41–47 (2017)
10. Mihai, S., Filip, V.: 3D modeling and performing of orthopedic implants by material deposition rapid prototyping. Rom. Rev. Precis. Mech. Opt. Mechatron. **41**, 128–131 (2012). (ISSN 1584–5982)
11. Mihai, S.: Orthopaedic implants and prostheses. Conventional and unconventional processing technologies. In: International Conference on Architecture, Civil & Mechanical Engineering (ICACME'2012), pp. 132–136 (2012). ISBN 978-93-82242-15-4
12. Mihai, S., Filip, V.: Aspects of orthopedic implants and prostheses. Mater. Process. Technol. Sci. Bull. VALAHIA Univ. Mater. Mech. **7**, 101–106 (2012). (ISSN 1844-1076)
13. Cintra, F.F., Yepéz, A.K., Rasga, M.G.S., Abagge, M., Alencar, P.G.C.: Tibial component in revision of total knee arthroplasty: Compararison between cement and hybrid fixation. Sociedade Brasileira de Ortopedia e Traumatologia **46**(5), 585–590 (2011)
14. Williams, A., Logan, M.: Understanding Tibio-Femoral motion. Knee **11**, 81–88 (2004)
15. Robalo, T.: Analysis of bone remodeling in the tibia after total knee prosthesis. Instituto Superior Tecnico – Universidade Tecnica de Lisboa (2011)
16. Badita, L.L., Gheorghe, G.: Orthopaedic prosthetic systems topographically characterized at nanometric scale. Adv. Mater. Res. **837**, 675–681 (2014)
17. Florescu, V., Căpitanu, L., Bădiţă, L.L., Filip, V.: A novel engineering spherical bearing, with potential application for a hip implant. J. Mech. Eng. Autom. (JMEA) **6**(5), 217–226 (2016). (ISSN 2159-5275)
18. Florescu, V., Capitanu, L., Badita, L.L., Filip, V.: A surprising tribological validation: metal on metal total hip prosthesis with rolling friction. Int. J. Appl. Innov. Eng. Manag. (IJAIEM) **5**(7), 025–038 (2016). (ISSN 2319–4847)

Ankle-Knee Rehabilitation System

Amin S. Abdullah Abdullah(✉), Cristian Gabriel Alionte, and Constantin Nitu

University "Politehnica" of Bucharest, 313, Spl. Independentei, 060042 Bucharest, Romania
almuhandes1979@yahoo.com, cristian.alionte@upb.ro

Abstract. In this paper is present a new type of a rehabilitation system for both knee and ankle that integrate two treatment types. The system allows a continuous passive motion consisting of two rotations of the leg segments, one of the knee articulation and one of the ankle joint, allowing the functional rehabilitation of the individuals that have an impairment of walking due to a disease or an accident. This is one novelty of the system because all the system that are on the market now consider to one articulation that can be knee or the ankle. In addition, the system allows the application of a mechanical vibration at the lower level of the leg, which have a beneficial action on the bones, muscle, and nerves, which is a well-known treatment.

Keywords: Vibration rehabilitation system · Functional rehabilitation Modelling · FEM analysis

1 Introduction

It is well known that, for the realization of a mechatronic gait simulation system must be considered several input parameters those which are as closest possible with the real ones owned by the patient. The parameters can be obtained from measuring using non-invasive methods, usually optical ones [1], because these parameters are individualized for each individual due to various factors:

- Body features: height and weight;
- Anatomy of the person: length of body segments, muscle tone;
- Psychosomatic features: personality, activism, type of life (sedentary, sporting), etc.;
- The gait ontogenesis because each person during childhood acquire two-leg and gait body stability mechanisms using individualized systems and methods.
- Gait mechanism is into a continuously transformation which takes place throughout the life span.
- Gait parameters usually are obtain using anthropogenesis measurement as function of averaging different actual measurement and therefore it cannot be said that gait is correct unless it adversely affects other components of the human body including gait.

Using the before mentioned facts can be conclude that when it is desired to develop a medical rehabilitation system for walking, it is important to consider the individual kinetic parameters of the patient. These can be done through the development and

G. I. Gheorghe (Ed.): ICOMECYME 2018, LNNS 48, pp. 166–172, 2019.
https://doi.org/10.1007/978-3-319-96358-7_17

implementation of a simpler walking analysis system that can be adapted to any patient and does not lead to a change in the usual course of the patient. In literature are several studies regarding human gait devices starting from Leonardo da Vinci, Galileo, Lagrange and Bernoulli [2] that have an interest in human gait until now. One of the first modern devices for gait assistance was developed by Hughes [3], Townsend [5] and Vukobratovic [4] and exist more sophisticated means for human help [6].

In this article is presented a system, which was developed within the framework of the dissertation thesis of Abdullah [7], which aimed at researching the design and realization of a biomechanical system for the medical rehabilitation of the inferior limb by a continuous passive movement of the foot. Consideration has been given to balancing several factors in the design process, including cost and simplicity of design.

2 The Rehabilitation System

Continuous passive motion mechanisms are used for long-term by persons, which are bed immobilized, or after surgery or during the early postoperative period. Continuous passive motion mechanisms are available for many joints of the body, knee, ankle, jaw, hip, elbow, shoulder and finger joints. The study, presented in this article, analyses ensemble knee-ankle only.

A continuous passive motion mechanism was designed with six mobile elements and eight joints as can be seen in Fig. 1. The continuous passive motion mechanism used two motors, one motor for knee motion and another for ankle.

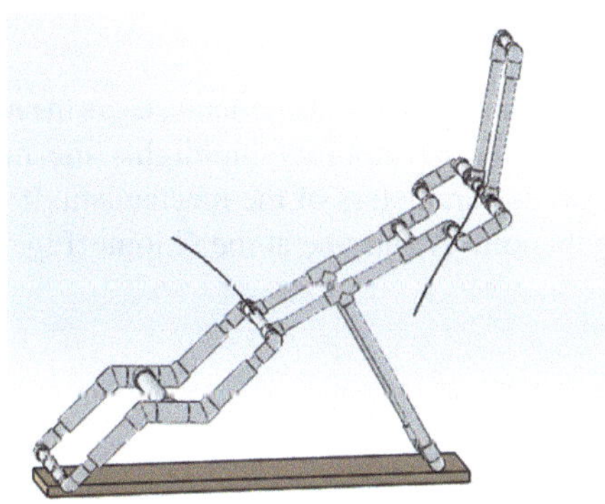

Fig. 1. The 3D model of the medical rehabilitation system of ankle and knee.

The action of the foot was realized by combined motion. Two elements can be adapted to variable length, so the study presents the mathematical model for each position. The passive motion was used to prevent ankle contracture and the vibration to lessen muscle tone, to strengthen muscle power and to prevent thrombosis. The system has three independent systems:

- system 1 for the movement tibia and fibula which will produce the rotation of the knee;

- system 2 for the movement of the forefoot which will produce the rotation of the ankle;
- system 3 for vibration excitation of the muscles: soleus, gastrocnemius, popliteus, plantaris, flexor digitorum longus, flexor halluces longus and tibialis posteriorus and bones the tibia and fibula.

The first system consists of one stepper motor, which is moving through a belt a frame on which the leg is fastened. The motor is a stepper one so the control is very easy to implement. The motion is transmitted to an L-frame, which has a linear guide on the bottom and is connected on a parallelogram system in the upper part. Also, this frame is support the system 2 which only have a rotational movement of the patient foot. The movement is generated using a second stepper motor. The vibration is produced using an exciter at the upper level of the L-frame and is transmitted through the structure to the muscles and bones of the leg.

3 Modeling and Simulation

Within the project we have developed a continuous rehabilitation medical structure consisting of six mobile elements and eight joints. We used two engines to move the structure, one at the knee and one at the heel that will generate a combined motion. The degree of mobility of the whole system can be described by Eq. 1:

$$M = 6n - \sum_{m=1}^{5} m * C_m \tag{1}$$

Where n represents the number of mobile elements, C_m is the number of pair elements with m constraints. The degree of mobility highlights the number of independent elements that define the status parameters of the mechanism. It will be considered two distinct solutions, one at the joint A and one at the G joint (Fig. 2).

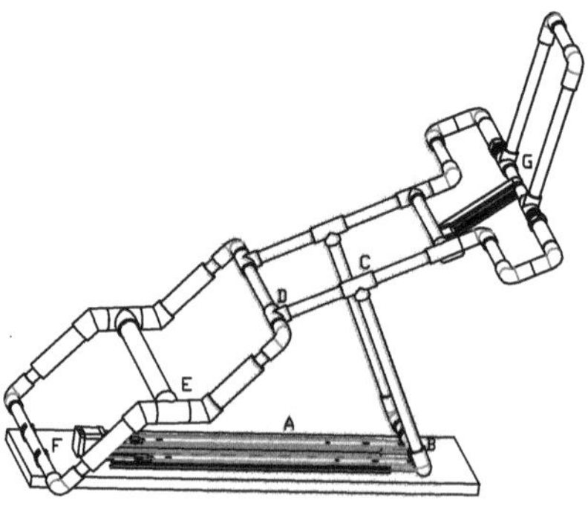

Fig. 2. The joints and elements of the medical rehabilitation system.

There are 2 elements DE and EF to adjust the length of the D-joint so we need to study the mathematical model that takes into account the two positions. Points G, E, C are chinking. We will identify vectors of the contour line positions associated with the kinematic chain, i.e., ABCDEFA, whichever is more oriented and positioned as far as possible from the point of view of the angle. These contour lines can be rewritten as vector:

$$\overline{AB} + \overline{BC} + \overline{CD} = \overline{DE} + \overline{EF} + \overline{FA} \tag{2}$$

The knee mathematical model is based on scalar equations obtained from vector equations. For a geometric and kinematic analysis, the following dimensions of the elements were considered. AB = 0.079[m]; BC = 0.328[m]; CD = 0.162[m]; DE = 0.244[m]; EF = 0.187[m]; FA = 0.3 … 0.047 [m].

To determine equation solutions, the PTC Mathcad Express Free [8] program was used based on a congruent program was used based on a congruent approximation method and the least squares method to solve the computational system. Several values were given for angular displacements, but only those values for which they allow continuous movement were selected. In Fig. 3 are presented the simulations results at the D joint (knee level) and at G Joint (ankle level) considering the displacement at the knee level. The simulation is taken into account the origin of the coordinate system that is at F joint.

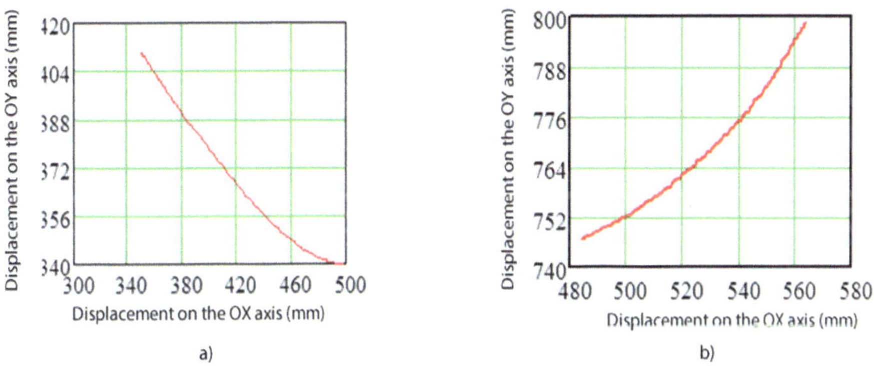

Fig. 3. The joints and elements of the medical rehabilitation system.

4 Results and Conclusions

The first analyses are take into account only the mechanical stresses that actioned on the leg support bars independently. The first simulation is done with the support platform fixed and a force of 10 N which is acting on the ankle sustain bar. The system has a minimal deformation of 0.006 mm and a maximum stress on the 1st frame vertical bars (Fig. 4).

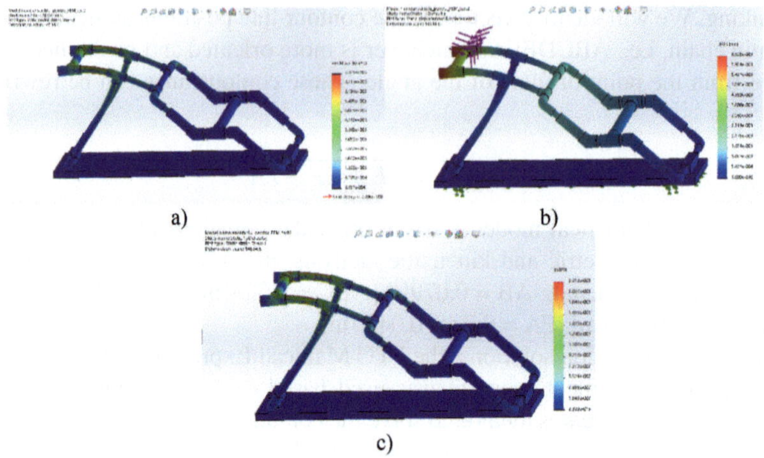

Fig. 4. FEM analysis results for a mechanical force of 10 N on the ankle support bar: (a) mechanical stress (b) deformation and (c) strain.

The second simulation is done with the support platform fixed and a force of 10 N which is acting on the knee sustain bar. The system has a minimal deformation of 0.0017 mm and a maximum stress on the 1st frame vertical bars and on the 2nd frame (Fig. 5).

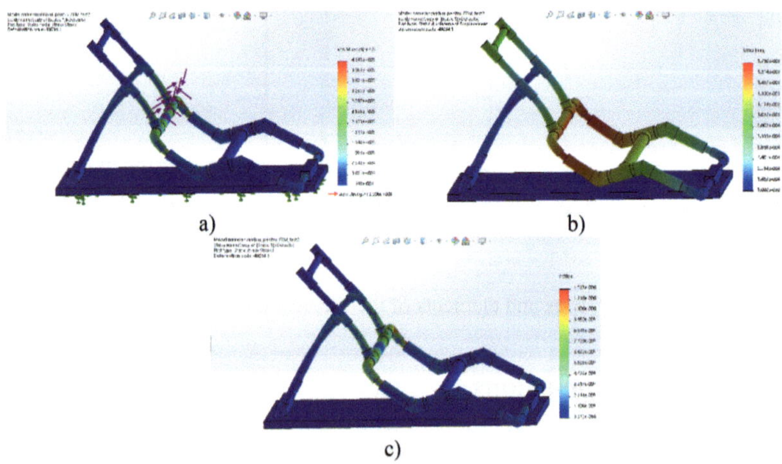

Fig. 5. FEM analysis results for a mechanical force of 10 N on the knee support bar: (a) mechanical stress (b) deformation and (c) strain.

The third simulation is done with the support platform fixed and a force of 10 N which is acting on the thigh sustain bar. The system has a minimal deformation of 0.09 mm and a maximum stress on the 2nd frame (Fig. 6).

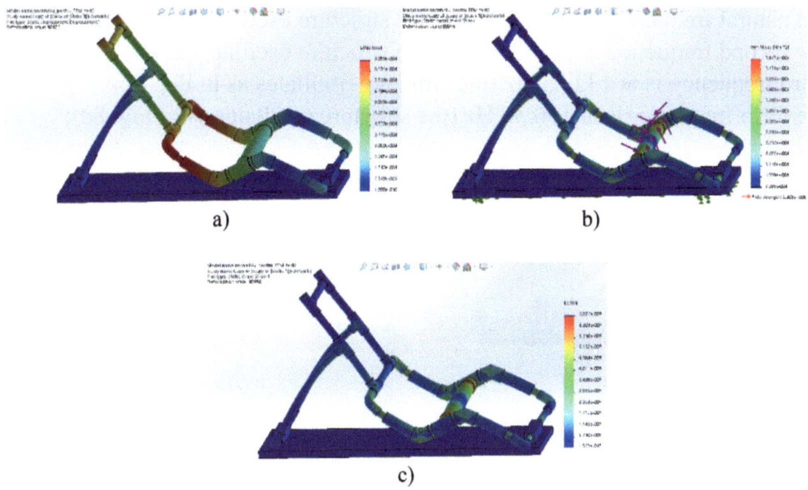

Fig. 6. FEM analysis results for a mechanical force of 10 N on the thigh support bar (a) mechanical stress (b) deformation (c) and strain.

The forth simulation is done with the support platform fixed and a force of 10 N which is acting on all three sustain bar. The system has a minimal deformation of 0.014 mm and a maximum stress on the 2nd frame (Fig. 7).

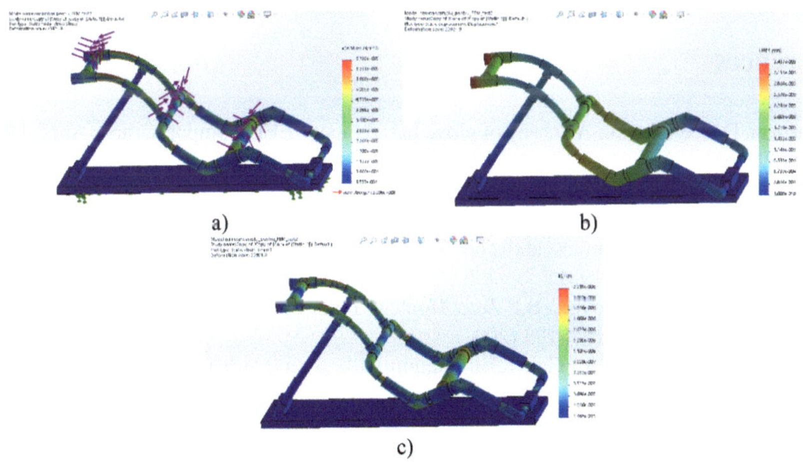

Fig. 7. FEM analysis results for a mechanical force of 10 N on all three support bar: (a) mechanical stress (b) deformation (c) and strain.

Continuous passive motion mechanism in combination with vibration reduced the muscle tone in the shank muscles and seem to be a safe and a feasible method to support ankle mobilization. Therefore it was made a modal frequency analysis to determine the natural frequency and 1st, 2nd, 3rd, 4th and 5th harmonics:

- the natural frequency is at 61.276 Hz (the structure oscillates as in Fig. 8a);
- the second frequency is at 117.21 Hz (the structure oscillates as in Fig. 8b);
- third frequency is at 132.05 Hz (the structure oscillates as in Fig. 8c);
- the forth frequency is at 156.36 Hz (the structure oscillates as in Fig. 8d);

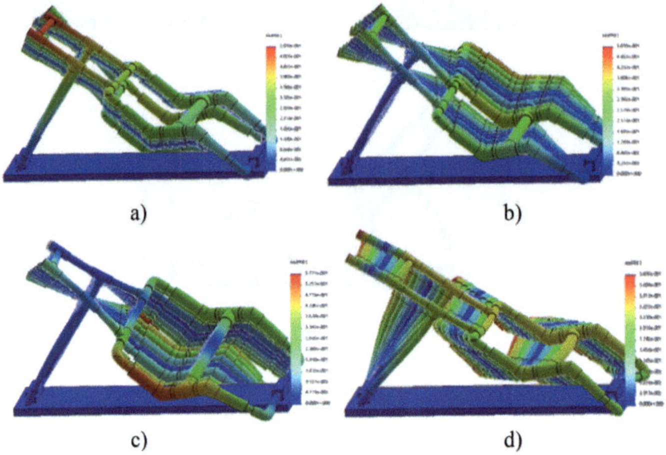

Fig. 8. Modal analysis: (a) the vibration of the structure at the frequency of 61.276 Hz; (b) the vibration of the structure at the frequency of 117.21 Hz; (c) the vibration of the structure at the frequency of 132.05 Hz; (d) the vibration of the structure at the frequency of 61.276 Hz.

References

1. Sturman, D., Zeltzer, D.: A survey of glove based input. IEEE Comput. Graph. Appl. **14**, 30–39 (1994)
2. Winter, D.A.: Biomechanics and Motor Control of Human Movement. Wiley, Canada (2009)
3. Hughes, J.: Powered lower limb orthotics in paraplegia. Paraplegia **9**, 191–193 (1972)
4. Townsend, M.A.: Dynamics and coordination of torso motion in human locomotion. J. Biomech. **14**, 727–738 (1981)
5. Vukobratovic, M., Borovac, B.: Zero-Moment Point-thirty five years of its life. Int. J. Humanoid Robot. **1**(1), 157–173 (2004)
6. Banala, S.K., Agrawal, S.: Gait rehabilitation with an active leg orthosis. In: Proceedings ASME Mechanisms and Robotics Conference (2005)
7. Abdullah, A.S.A.: Design and construction of a biomechanical system for medical rehabilitation. M.Sc., Dissertation Precision Mechanics, University "Politehnica" of Bucharest (2017)
8. PTC Mathcad Homepage: https://www.ptc.com/en/products/mathcad-express-free-download. Last accessed 2017/10/14

A Review of Fault Diagnosis in Mechatronics Systems

Daniel Cordoneanu(✉) and Constantin Niţu

Department of Mechatronics and Precision Mechanics, University "Politehnica" of Bucharest, 313, Splaiul Independenţei, Bucharest 060042, Romania
daniel.cordoneanu@stud.mec.upb.ro

Abstract. Fault diagnosis gets increasing importance in today's production environment and with the advancements in the field of artificial intelligence, researchers look for new ways to keep a system away from faults, that would interrupt the production. In recent years, many papers were written regarding this subject especially regarding predictive maintenance and fault diagnosis. This paper presents recent works that expose new methods for intelligent fault diagnosis. This step is important for future research in order to have a better understanding of state of the art algorithms and look for ways to improve the existing fault diagnosis approaches. The focus will be on electrical systems and actuators and manipulating systems (robots), production experience showing that the mechanical parts are the most exposed to production-ending faults. That's why most of the observer systems are using vibrations as the main data for their algorithms, but also other measurements can provide useful information about the condition of a system.

Keywords: Fault diagnosis · Predictive maintenance · Mechatronics systems
Artificial intelligence

1 Introduction

Mechatronics systems cover a wide area in today's production machines. Mechanical parts and electronics are fused together and now, with the Internet of Things (IOT), intelligent control algorithms can operate and control the machines remotely, having all the necessary power of computing at their disposal. However, we can split the mechatronics systems into multiple types:

- actuators – electrical motors, pneumatic and hydraulic motors, etc.
- mechanical drives – gearboxes, belt pulleys, worm gears, etc.
- sensors and data acquisition systems – force sensors, accelerometers, temperature sensors, etc.

Further, methods of condition monitoring and fault diagnosis for each of these subsystems will be presented, and then review ways of monitoring them as a whole, in complex systems.

G. I. Gheorghe (Ed.): ICOMECYME 2018, LNNS 48, pp. 173–184, 2019.
https://doi.org/10.1007/978-3-319-96358-7_18

The methods are often divided into multiple categories, as Mouzakitis describes the fault diagnosis systems in [1]:

- model-based methods
- data-based methods
- hardware-based methods

Further it will be presented different combinations of the methods used for the systems mentioned before. Firstly, a better differentiation between these methods is graphically shown. In Fig. 1, there are presented some of the algorithms that are used to implement the above-mentioned methods. These algorithms are well described by Mouzakitis in his article [1] and some of them will be further discussed in this paper, based on fault diagnosis examples for certain systems.

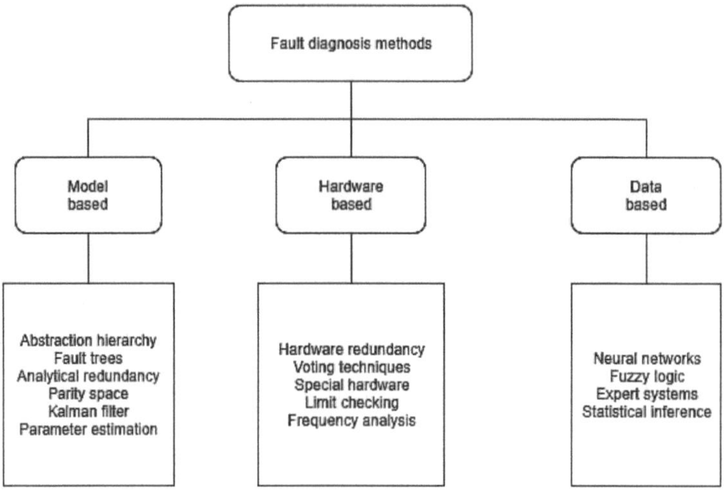

Fig. 1. Fault diagnosis methods and algorithms for implementation

Figures 2 and 3 show the main difference between model-based and data-based methods.

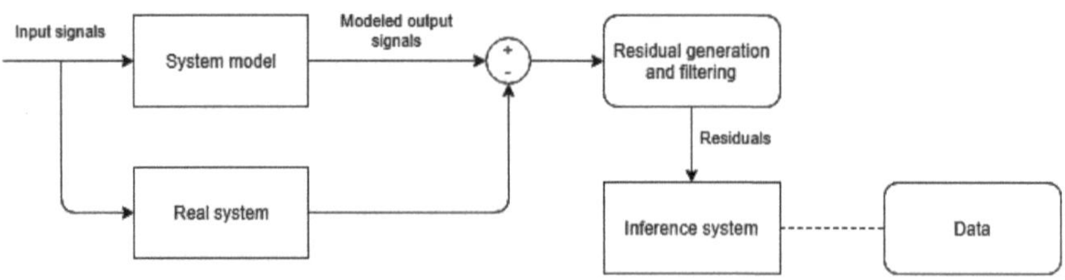

Fig. 2. Simple diagram of a model-based fault diagnosis system

As it can be seen in the two figures, they are pretty much the same at a macro system level, but the main difference consists of how the residual information is generated. The model-based one needs no prior data to get an expected result, while the data-based system requires information on which an internal model of the system is built. The quality and quantity of the training data is very important (as it is for most machine learning algorithms) and also the system has to be properly designed, in order to avoid the training data over fit or under fit. In both cases, the expected output would be erroneous, leading to a bad fault diagnosis. On the other hand, if the model-based system is too simple and does not capture all the dynamics of the system, then the fault diagnosis system may not be able to detect all the faults that may appear. In both cases, the inference system can be an artificial intelligence system, which needs data to be trained as well. In data-based methods, the statistical methods can also be included. For example, particle filters algorithms are very well described in [2] by Verma et al. They put in balance the computational power that is needed in a simple model-based fault diagnosis method and in a stochastic method. The latter, although not being as accurate as the model-based method, is more efficient and takes less time to process the signals and extract residual features needed for the health monitoring algorithms.

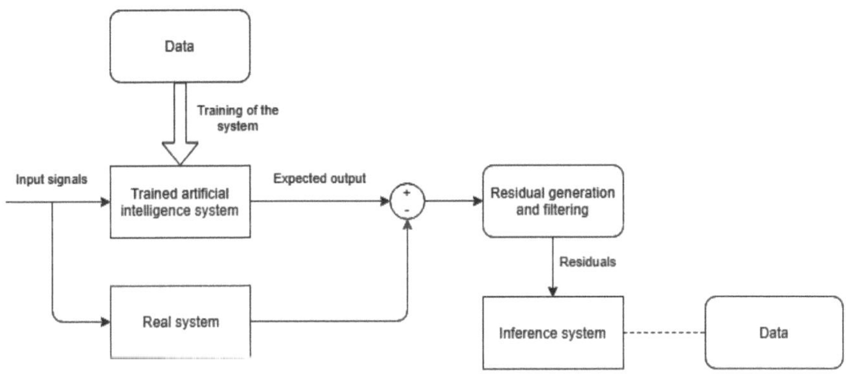

Fig. 3. Simple diagram of a data-based fault diagnosis system

An architecture example for fault diagnosis systems is given by Catterson and McArthur in [3], where a Multi-Agent system for the condition monitoring of a power transformer is presented. This one is organized by layers, that should be part of any fault diagnosis system, as COMMAS architecture, where they mention that each system agent can be represented by a different algorithms (a modular approach). These layers are, in fact, the following 4 tasks:

- data monitoring
- interpretation layer
- corroboration layer
- information layer

2 Fault Diagnosis for Actuating Systems

The actuating systems play a very important role in every dynamic system, because they are the "fuel" of the production systems.

Starting with a model-based fault diagnosis approach, Rolf Isermann presents some fault-detection methods based on the system model and then points out an application of these ones on a real system [4]. The methods that can be used are summarized as:

- process models and fault modelling
- fault-detection with parameter estimation
- fault-detection with observers
- fault-detection with signal models

The author's diagram of a model-based fault detection system is shown in Fig. 4.

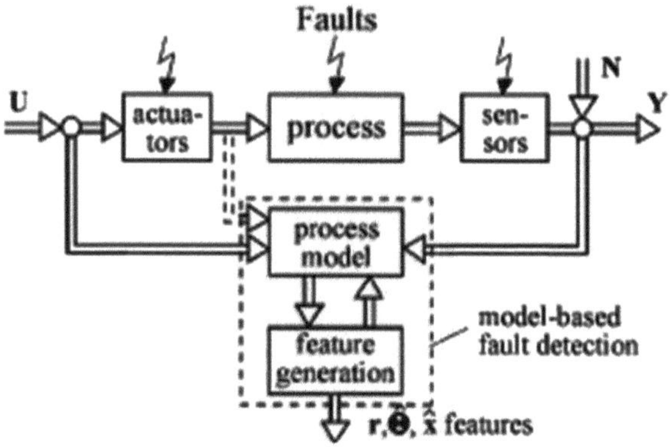

Fig. 4. Scheme of process model-based fault-detection [4]

In the article, the author presents an example of a fault diagnosis algorithm for a cabin pressure outflow valve actuator. It is not only an interesting case, but also safety related since the cabin pressure is vital for the passengers. There are two brushless direct current (BLDC) motors, one of them being a backup, if the other one has a fault. The fault-detection method is a combination of parameter estimation and parity equations. Having a detailed BLDC motor model for all three phases and based on the model as well as the available measurements, the needed parameters for generating residuals were estimated and the parity equations were created. The measured values are control voltage-U(t), armature current-I(t), the angular position of the shaft and of the flap. The last two are related by a gear ratio. The residuals allow a monitoring algorithm to detect and isolate the faults (in this case, a fuzzy logic system). Each residual represent a certain measure and based on these signals, the system can detect the part of the system that is faulty. The author's model is presented in Fig. 5.

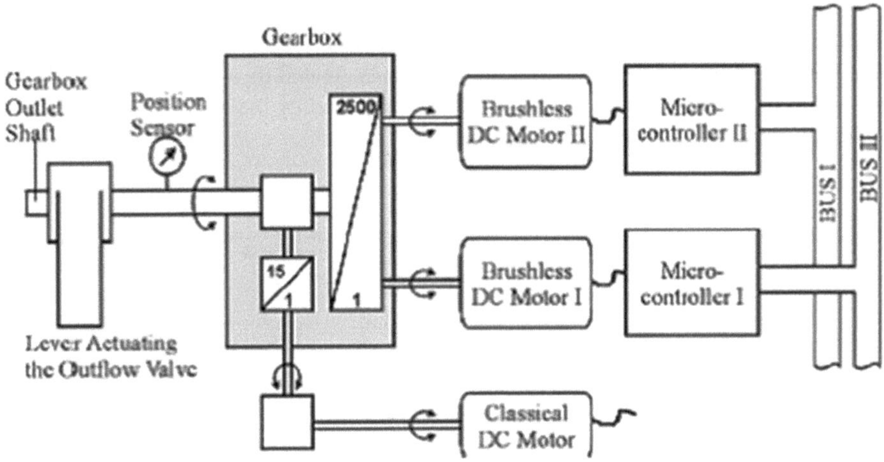

Fig. 5. Redundant DC motor drive system for the outflow valve [4]

In this example, it can be noticed that the electricity provides the information about the state of the system. A model-based solution gives the advantage that each part of the system is described and faults isolation is only a matter of comparing the modelled signal with the real one, thus obtaining residuals which can further be stored in a fault matrix. One disadvantage of this method is encountered when the system has non-linear dynamics and multiple components and consists of a lot of computational power consuming. Another drawback is the need for changing the model, if a part of the system is replaced with a different part which has the same output, but different behavior.

Another model-based fault diagnosis method is proposed by Cho et al. [5]. Their method is based on a stochastic model using only the current measured by a Hall sensor. The authors' algorithm is best described by the diagram in Fig. 6. The main advantage of this approach is that the probabilistic model is part of the system and the inputs that go into the real system do not need to be computed again through the system's model. Therefore this kind of method is very efficient, having a low computational load.

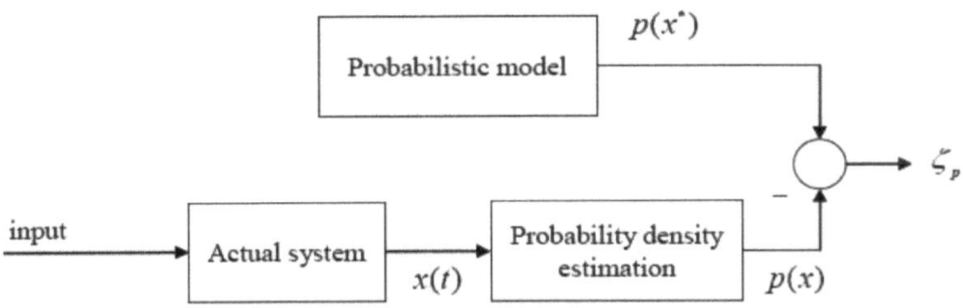

Fig. 6. Stochastic model based fault detection scheme [5]

To address a general situation where a model is hard to create, data-based solutions are the best fit. For instance, in [6], Su and Chang propose a fault diagnosis method based on vibration signal and on vibration residuals created by a neural network. It can be seen in Fig. 7 the model that the schematic proposed by the authors is more or less the same with the one illustrated in Fig. 3.

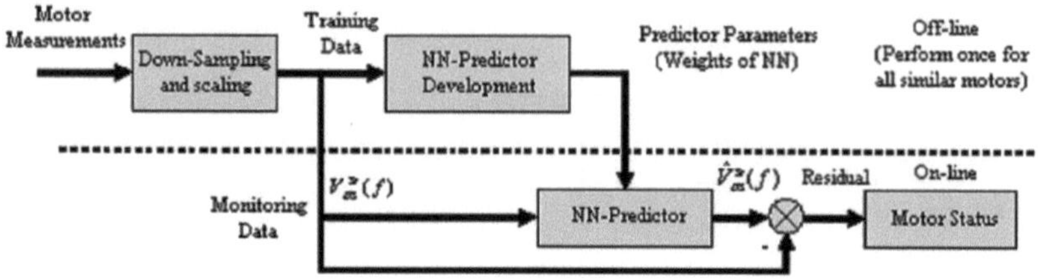

Fig. 7. Principle of analytical redundancy method using neural network modelling [6]

The system which is monitored is an induction machine, a widely spread electric actuator in the industry. An induction machine's model has many non-linear characteristics and to create a dynamic model of this system is time consuming and also takes a lot of computing power. Temperature, inverter variations, etc. are all factors which come into play when analyzing the vibrations signals that come from this type of machine. As the authors mention, vibrations signals are accepted as a good indicators of the stator, rotor and bearings condition.

Comparing this method with the first one, some differences have to be mentioned:

- model-based methods give a more clear indication about the source of the faults
- data-based methods are more versatile because they can capture non-linear dynamics without having to know specific parameters of the system
- both methods use in the end residuals as inference parameters on whether there is a fault in the system or not

An alternative for dynamic system efficient monitoring is based on vibration analysis. In [7] the authors state that "vibration analysis techniques are more effective in assessing a machine's health". In this paper, a neural network is used to create a data-based model for the electrical behavior of the machine. The data used is extracted from the machine vibration signals and they represent quasi-steady signals used to remove unwanted disturbances. Further, the neural network output is then used to create the analytical residuals on which the fault diagnosis inference is based on. Both in [6, 7], this type of fault diagnosis using artificial intelligence is up to the task of modelling properly a non-linear dynamic system and the output can be used then to assess the system's health condition.

3 Fault Diagnosis for Mechanical Drives

The motion transmission systems play an important role in every engineering system because they are the ones which transfer the power developed by the actuating systems. That's why fault diagnosis techniques have been developed for these systems as well. A challenge, however, for this type of systems is to create a general fault diagnosis system, which is not aimed to a certain type of components, but is focused on the kinetic energy transfer chain: actuator - mechanical drive - moving object. For example, the same type of motion can be achieved using a gear-rack transmission or a screw drive: they both convert a rotational motion to a linear one.

For example, in [8], Wang and Wong propose a model-based fault diagnosis technique for gear faults that relies on signal averaging technique. A model is firstly created, based on the vibration of the monitored gear in its healthy state. This model is then used as an error filter to compute the future-state of the vibration signal from the same gear. The authors' algorithm is presented in Fig. 8. As the diagram stops at the prediction error, the algorithm can be improved by using this error as data for a machine learning technique that would give more insight into what type of fault it is and, maybe, having the gears still working at a lower speed (having a fault tolerance).

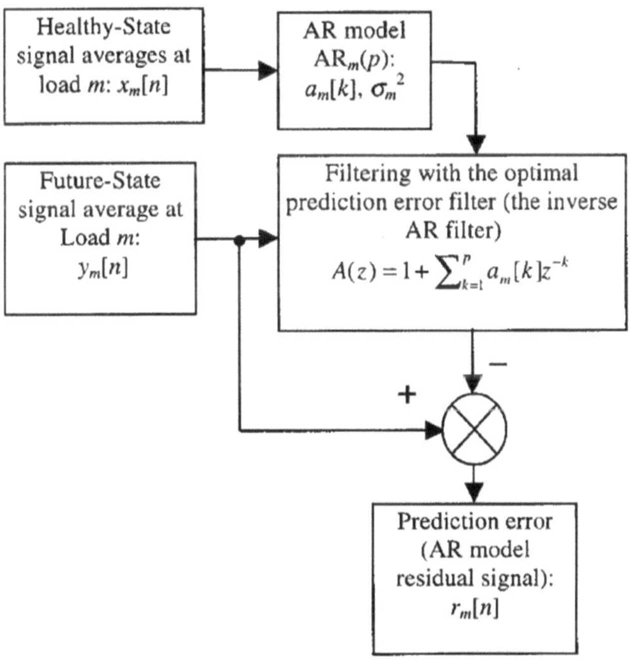

Fig. 8. Model-based fault diagnosis of a gear [8]

The previously described fault diagnosis methods were applied to a belt conveyor, in order to make a comparison of results. In [9], Ghoshal and Samanta present a method in which the belt conveyor system is observed and diagnosed using a model-based approach. This one makes use of bond graphs formalism to model the system, which is based on the power exchange between the ideal components of the dynamical systems. Mainly, there are power variables sources, transformers, storing and dissipative elements, characterized by lumped parameters. Moreover, this way of modeling is independent on physical domain, enabling unitary representation of the multidisciplinary systems. For example, the belt conveyor model, shown in Fig. 9, involves both electrical and mechanical domain.

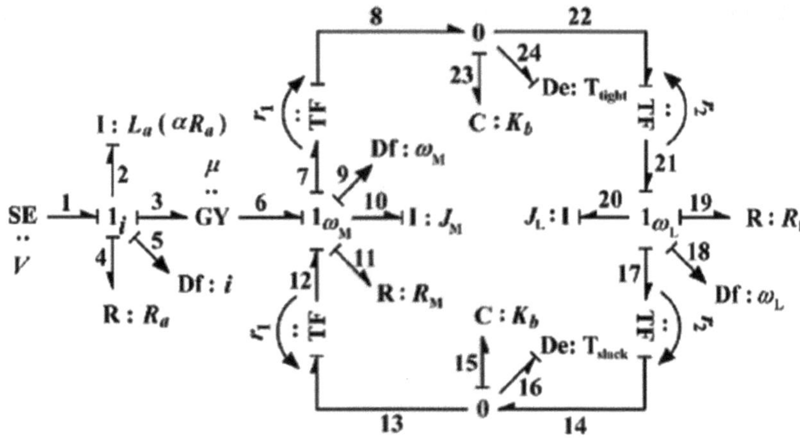

Fig. 9. Bond-graph model for a belt conveyor system [9]

Using this type of modelling, the authors have relied on Analytical Redundancy Relations (ARR), which are constraint relationships between inputs, outputs and known process parameters (or estimated). The numerical result of ARR called residual is then used to asses if there is a fault present or not in the system. Also, these residuals are used to isolate the faults by help of a fault signature matrix. Because in real time applications the residuals which result after processing the data from the sensors will never be zero (sensor noises, parameters uncertainties), a coherence vector is built for each residual that takes into account some threshold values. A fault is detected when one of the elements of the coherence vector is different to zero.

Having the ARRs in differential form (so someone can look backwards in time to see what happened) and getting the residuals in comparison with the fault signature matrix, it is possible to identify which element of the bond graph is responsible for the fault. This way, the corresponding real system component is identified. In this example, a simple belt drive with two pulleys is modelled and simulations are run for the cases when there are no faults and when there are faults. The losses due to damping or jamming of the pulley are measured in the energy-dissipative elements of type R (the damping coefficients of the two pulleys), which will further be observed and based on their values, assertions of different faults can be made. The sensors measure the motor current, the rotational speeds of the pulleys and the tensile forces in the two branches of the belt. This type of fault diagnosis is pretty straight forward and easy to implement, the hardest part of it being the modelling one and finding out the correct threshold values for the coherence vector.

In [10], Li et al. analyze a system that has a belt drive, this time the fault diagnosis method involving an artificial intelligence system similar to neural networks - support vector machine (SVM). The only collected data is from vibration in order to have the smallest number of sensors. This data is further processed to retrieve the wavelet packets (as the authors point out, this type of data is efficient for representing a rotary machine). The wavelet package decomposition (WPD) is based on wavelet analysis, giving an improved time-frequency resolution. Once the signal was filtered and the energy for each frequency band was obtained, this data can be used as input to the SVM system.

The SVM is an algorithm based on statistical methods that copes very well with nonlinear, finite-sample, high-dimensionality problems. This algorithm is efficiently used to classify the state of the conveyor belt and idlers. In this specific scenario, the authors point out that in real applications it's pointless to add sensors to each idler (and not cost-efficient), so they added accelerometers once every three idlers. The data that come from these sensors are then filtered out to WPD and represent the input of the SVM system. The output of this system will be provided to the management personnel that will decide on how to approach the presence of an early fault in one of the idlers.

The method is good for systems that require a limited number of sensors and in contrast to the model-based method, it's easier to use in systems that have a nonlinear behavior (which sometimes is very hard to model/estimate). Also, it can be noticed that the input for this fault diagnosis system in represented by only a signal (acceleration), while for the model-based approach, several input signals were employed. Additionally, this type of fault diagnosis is strongly dependent on the sensors output, so if there would be a fault in the sensors, the system would probably signalize false faults.

Going further, there is a method with which also sensors' health is monitored.

4 Fault Diagnosis for Sensors and Data Acquisition Systems

In [11], Ding et al. present an advanced way of monitoring sensors that play a significant role in 3 vehicle systems: ABS (anti-lock breaking system), ESP (electronic stability program) and TCS (traction control system). The fault diagnosis system uses a model-based method, in which the sensors models are developed and 3 observers are used, each one generating two residual signals. In this work, the observers are built based on three time-discrete sensor models, which are suited to the vehicle dynamic system. The first residual signals are used for estimating model uncertainties and they are called Control Signals. The latter residual signals are called Detection Signals and are used for indicating the sensor faults.

This method shows the advantages that a model-based fault diagnosis system can provide if the model is thoroughly developed and can detect sensor faults and even estimate the faulty sensor signal. For this type of systems, which other fault detection systems rely on, this is the best way to ensure a health monitoring, given the fact that if a signal is not correct, the entire system control can go wrong.

In [12], Jan et al. describe an intelligent method of monitoring the sensors using Support Vector Machines to classify faults that can appear in a temperature-to-voltage converter. They identify five faults that can appear in the sensor and train the system to classify the data into one of these faults or into a healthy system. Comparing this algorithm with the previous one, one can state that the latter is easier to implement, the hardest part being the training of the classifier. On the other hand, this system can only tell if there is a fault in the sensor, not being able to estimate a correct sensor signal and it's more prone to error than the model-based one. Of course, based on the type of sensor, each method has its benefits or downfalls, but based on the type of information that the sensor offers (life-critical data or just system data) the model-based system should be used over the artificial intelligence.

Since all the important subsystems that are part of a mechatronics system were covered, further it will be presented and discussed a fault diagnosis approach that tries to monitor the health of the entire system.

5 Fault Diagnosis for a Complex System (Robot)

Firstly, it is a look into a stochastic model-based method proposed by Mendoza et al. in [13], where they use the healthy robot model to assess whether there is a fault or not in a mobile robot. The method relies on three redundant information sources that can be used by an inference system. However, this article only presents a high level approach of a more complex task of identifying and detecting faults in a very complex system.

Another fault diagnosis approach is proposed by Liu and Coghill in [14], where a complex system is monitored using a model-based technique. This online robotic fault diagnosis approach is called "First Priority Diagnostic Engine" (FPDE). The proposed method presents a health monitoring system that is seen as a separated package from the robot itself and which detects sensor faults, model faults and robotic faults based on a first-priority parameter that is chosen for the robot and which is considered the principal parameter to be monitored.

The authors explain what robotic faults are. They can be described as orientation faults or translational faults, which can be caused by different components that are faulty. The system has an underlying system which is called "Component-based robotic reasoning", which narrows down the fault at component level. They also present the reasoning algorithm, which is based on the interval filter (which is a system that extracts characteristic values of robot parameters), and which can actually categorize the faults "into behavior faults and modeling faults, based on sensor fault detection". There are two other underlying systems: one is the bounds generator, which creates different bounds based on the robotic model for the different variables that are monitored; the other system is the fault reaction which decides what actions the robot should take, based on the discovered faults. This approach encapsulates artificial intelligence (interval filter) and model-based fault diagnosis and it can also be used for various mechatronics systems, while it's very useful for checking a system's model as well. However, there are faults that can be diagnosed earlier, based on vibration patterns, before a certain monitored parameter goes out of bounds (for example the manipulator's trajectory).

In [15] Anand et al. propose a hybrid intelligent approach for diagnosing a robot's manipulator. The authors use neural networks to model the robot's model nonlinearities and then use fuzzy logic to make an inference for detecting there is a fault or not in the robot's dynamics. This system also proposes a fault tolerance capability, which would allow the robot to work even if it's faulty. This way, the down-time is reduced and although there are faults in the robot, the production is not stopped. This article is very interesting as it captures very well the problems that have to be solved by a robust fault diagnosis system and it clearly states that having artificial intelligence algorithms makes the task much easier for online monitoring in complex systems. Figure 10 is the scheme that describes quite accurately a predictive maintenance system for a complex system on a high level, according to the above presented algorithm.

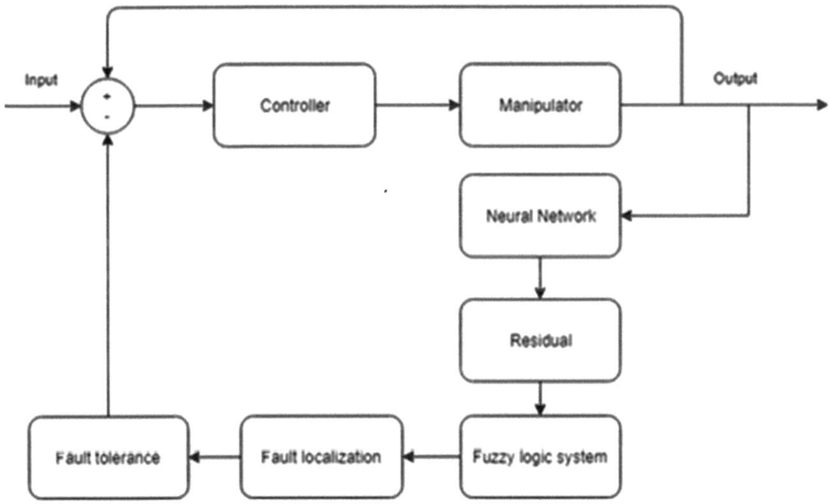

Fig. 10. Diagram for a fault diagnosis system for a robot [15]

6 Future Work

Starting from the concepts defined in the above presented works, a fault diagnosis system for a discrete model of a mechatronics system, that is generated by an artificial intelligence system will be developed. This one could be used for a general system, that is treated as a combination of damped oscillators.

References

1. Mouzakitis, A.: Classification of fault diagnosis methods for control systems. Meas. Control **46**(10), 303–308 (2013)
2. Verma, V., Gordon, G., Simmons, R., Thrun, S.: Real-time fault diagnosis. IEEE Robot. Autom. Mag. **11**, 56–66 (2004)
3. Catterson, V.M., McArthur, S.D.J.: The industrialization of a multi-agent system for power transformer condition monitoring. In: Applications and Innovations in Intelligent Systems XII, pp. 165–178 (2002)
4. Isermann, R.: Model-based fault-detection and diagnosis – status and applications. Annual Reviews in Control **29**, 71–85 (2005)
5. Cho, H.C., Kim, K.S., Song, C.H., Lee, Y.J., Lee, K.S.: Online fault detection and diagnosis algorithm based on probabilistic model for induction machines. In: SICE Annual Conference (2008)
6. Su, H., Chong, K.T.: Induction machine condition monitoring using neural network modelling. IEEE Trans. Ind. Electron. **54**(1), 241–249 (2007)
7. Su, H., Xi, W., Chong, K.T.: Vibration signal analysis for electrical fault detection of induction machine using neural networks. In: International Symposium on Information Technology Convergence (2007)
8. Wang, W., Wong, A.K.: Autoregressive model-based gear fault diagnosis. J. Vibr. Acoust. **124**, 173 (2002)
9. Ghoshal, S.K., Samanta, S.: Model based fault diagnosis of a belt conveyor through parameter estimation
10. Li, W., Wang, Z., Zhu, Z., Zhou, G., Chen, G.: Design of online monitoring and fault diagnosis system for belt conveyors based on wavelet packet decomposition and support vector machines. Adv. Mech. Eng. **5**, 797183 (2013)
11. Ding, S.X., Schneider, S., Ding, E.L., Rehm, A.: Advanced model-based diagnosis of sensors faults in vehicle dynamics control systems. In: 16th Triennial World Congress, Prague, Czech Republic (2005)
12. Jan, S.U., Lee, Y.-D., Shin, J., Koo, I.: Sensor fault classification based on support vector machine and statistical time-domain features. IEEE Access **5**, 8682–8690 (2017)
13. Mendoza, J.P., Veloso, M., Simmons, R.: Mobile robot fault diagnosis based on redundant information statistics
14. Liu, Honghai, Coghill, George M.: A model-based approach to robot fault diagnosis. Knowl. Based Syst. **18**, 225–233 (2005)
15. Anand, M.D., Selvaraj, T., Kumanan, S.: Fault detection and fault tolerance methods for industrial robot manipulators based on hybrid approach. Adv. Prod. Eng. Manag. **7**, 225–226 (2012)

Pipe Cracks Detection Methods – A Review

Ahmed Sachit Hashim(✉), Bogdan Grămescu, and Constantin Niţu

University Politehnica of Bucharest, 313, Splaiul Independenţei, 060042 Bucharest, Romania
ahmedhashim774@yahoo.com

Abstract. A comparison between the most important methods for crack detection during external inspection of the pipes is presented. It is meant to choose the best suited one for robotic devices, aimed to perform the pipelines monitoring, because they are often under the action of the environmental stress factors, like high-temperature variation, especially met in the Middle East countries. As a native from the Republic of Iraq, the first author currently works at development of a robot for external inspection of pipes, within his Ph.D. thesis, in order to provide full care to the pipelines, because they are non-replaceable. This way, their performance and reliability can improved and the production stopping is prevented.

Keywords: Crack detection · Camera detection · Acoustic detection
Magnetic field

1 Introduction

Crack is one of the most common defects occurring in the metal parts, with particular consequences to fluid transport pipes. The cause of the cracks generation is the exposure of the metal to mechanical and thermal stress, resulting in a defect consisting in the separation of the internal crystals, due to fatigue. For pipes, this indicates there is a turbulent pressure inside, during operation, which leads to a slight change in their shape, resulting in gradual weakness and the appearance of cracks, which can grow larger over time.

The most important methods used to detect the pipes cracks, which can fit to a mobile robot for external pipe inspection are: computer vision systems (camera), magnetic field measurements and acoustic detection (microphone).

2 Cracks Detection by Use of a Camera System

Due to the high importance of this type of detection, it has attracted the attention of many scientists and researchers in the field, mostly for crack detection in the pipes that are subject to high temperature [1]. Image analysis techniques were also used to detect welding defects in pipelines [2], as recently proposed thermal image analysis system for non-destructive tests (NDT), during thermal stress tests [3]. The use of a highly sensitive system to detect small cracks (about 500 μm) [4] proved that detection of cracks based

G. I. Gheorghe (Ed.): ICOMECYME 2018, LNNS 48, pp. 185–193, 2019.
https://doi.org/10.1007/978-3-319-96358-7_19

on thermal imaging in excitation methods depends on eddy currents [5]. Pulsed phase thermography and derivative techniques for cracks detection were also developed [6], while a laser technology for "active thermal imaging of the spot plane" was used [7]. This one is based on the local excitement of the point where the laser beam from the camera falls on the inspected pipe, similar to the analysis method used in the project ThermoBot [8]. An example of cracks detection scheme by use of camera vision systems is presented in Fig. 1.

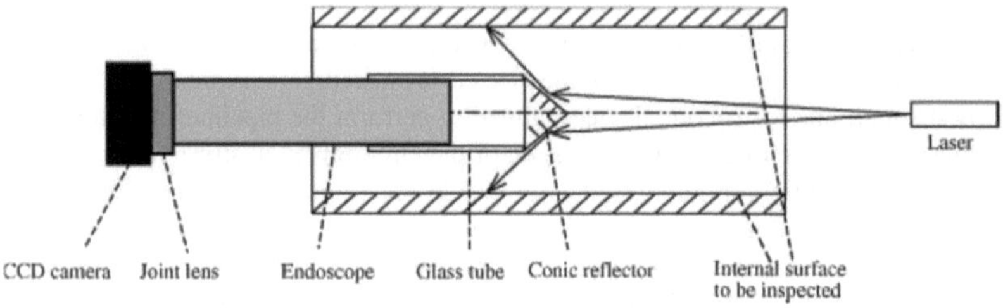

Fig. 1. An industrial endoscope used for building a laser profilometry platform [18].

While the camera is working on the test of cracks in the pipes, a white spot surrounded by an area of low gray appears due to heat spread from the hotspot. The algorithm used to detect cracking of metal parts depends on the analysis of the area, where the heat is spread. The heat zone analysis is performed in three steps: hotspots detection, analysis of the radius of the pipe and tangential gradient analysis.

As a first step, the hotspot should be detected on the pipe in the crack area. Because the laser spot is part of the top-temperature image, a simple algorithm can detect hotspots. A dynamic threshold is applied, depending on the lower and higher temperature values in the image, firstly to obtain a binary image, in which the laser position is the only white element. In the end, the center point is evaluated as the center of the mass of the resulting shape. Once the shape and location of the laser are available, it is possible to analyze the surrounding area to conduct detection of cracks in any smooth metal.

The first analysis involving the heat flow zone is performed on the radial gradient. While the normal regression factor takes into account the pixel difference only along the horizontal and vertical axes, the radial version takes into account multiple directions intersecting at a central point, corresponding to the hot point. Midtones are usually evaluated by comparing points aligned along a given direction, intersecting with the center point. However, when evaluated in the discrete area of the image, one important side effect should be considered. The number of pixels in a given distance to the center is not constant, but depends on the distance itself. This is important in making comparisons. For example, in [18] it is suggested another approach instead of comparing pixel values, focusing on variable-sized image areas, which is based on distance to the entrance. Another slightly different approach to the problem is based on moving from the outside to the center. Consider the part of the image in the Fig. 2 (left). The center is green in the lower left corner. Per pixel, where the vector connecting it to the center appears, and its color is red if it crosses only one pixel, and blue if the pixels are intersecting two. It is important to separate these two cases in the form of pixels with a red vector that can be compared to another pixel (referred to by the vector), as in any other grading algorithm. Blue vector pixels need to be compared to more than one other pixel, the two types that intersect with the center vector, as shown in the Fig. 2 in the middle left, where each pixel is compared to more than two pixels.

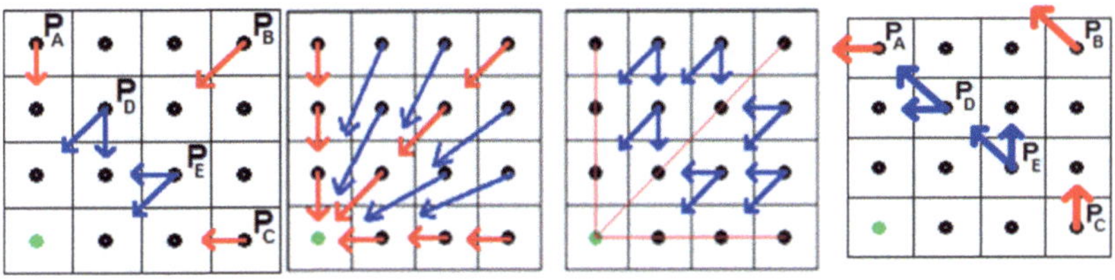

Fig. 2. Four types about pixel comparison of the tangential gradient algorithm [18].

Consider the parts of the image with the increasing area described in [18]. The use of the method described above can be identified in four different cases, which are summarized in Fig. 2. The (PA, PB, and PC) pixels are connected to the center on a vertical, diagonal and horizontal direction respectively; therefore, the radial gradient is evaluated by comparing it to a single pixel, which is indicated by the vector. In the case of (PD), the line connected to the center of the image contains a stronger vertical component, therefore, the (PD) must be compared with the pixels placed along the left and right direction, (PE) is bright, so the values to be assumed are the following:

$$\begin{aligned} P'_A &= P(X_A, Y_A + 1) - P_A \\ P'_B &= P(X_B - 1, Y_B + 1) - P_B \\ P'_C &= P(X_C - 1, Y_C) - P_C \\ P'_D &= f(P(X_D - 1, Y_D + 1), P(X_D, Y_D + 1) - P_D \\ P'_E &= f(P(X_E - 1, Y_E), P(X_E - 1, Y_E + 1) - P_E \end{aligned} \tag{1}$$

where $P(X, Y)$ is the point of coordinates (X, Y), and $f(\cdot)$ is a function that controls the different weight of adjacent pixels.

The gradient evaluated in the area around the laser one is not homogenous and is higher in locations close to the spot, where it can be easily understood by considering heat transfer within the metal parts. From the point of view of image processing, this breaks the homogeneity of the area being inspected and, therefore, must be differentiated. To balance this effect, a correction called radial gradient equalization (RGE) was proposed, which amplifies the gradient that depends on the distance to the center point of the laser region. The equation is applied in a region surrounding the laser location until the maximum length specified by the parameter is reached. Amplification is defined as a function of the distance to the center point and increases the linearity of the value until it reaches a certain value where it is accessed remotely, and then acquires saturation:

$$RGE(l) = \begin{cases} G_{MIN} + \alpha l & l < L_{LIM} \\ G_{MAX} & L_{LIM} < l < L_{MAX} \end{cases} \tag{2}$$

The results obtained using RGE can be seen in the Fig. 3, where the cracks are seen in the bottom edge after applying the equation in the direction right. The original picture, can be seen in the direction left.

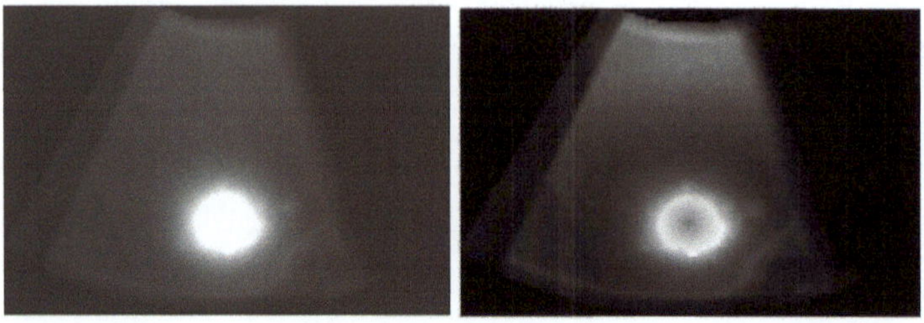

Fig. 3. Comparison between the images before and after use of the radial gradient equalization equation (RGE) [18].

The proposed algorithm is designed to detect cracks by analyzing radial gradient images and has best results when adopting the Eq. (2). To detect cracks, the gradient image is divided into parts of small patches that are analyzed separately. Each correction is numbered using an adaptive threshold, based on the average pixel value. Then, some image enhancement functions are applied, and finally, the cracks are detected by setting the features in the large size image, ignoring any image containing the spot of the laser. The final detection algorithm for cracking is fairly simple, because it works on images that are strongly strengthened at the lower level. An example of detecting cracks can be seen in the Fig. 4, where a binarized image is displayed in the left hand side with the final result in the right hand side.

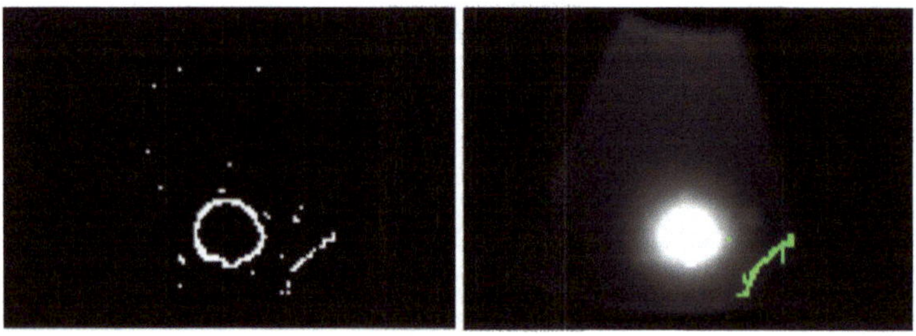

Fig. 4. An example of image uniformity (left) and of the end result (right) of the used algorithm [18].

The tests of this algorithm included the detection of cracks in single images and full sequences. It is ideal for the case of smooth metal parts with circular shapes such as pipes. The tests showed a good performance of the algorithm, which was able to detect cracks in the area surrounding the laser fall site. This means that the cracks are detected when the mineral areas lie farther than the laser spot. The crack was very small, with a length of 8.36 mm and only 120 μm [18].

3 Cracks Detection by Use of a Magnetic Field

Magnetic field leakage (MFL) inspection is a non-destructive test (NDT) used to detect possible pipelines defects such as cracks, corrosion, drilling, scratches, pits, voids, etc. It is a wide-ranging method for detecting cracks in both axial and transverse directions. MFL method is a more powerful one compared to the eddy currents method in terms of variation of magnetic properties. The pulsed magnetic field leakage (PMFL) sensor system is designed to predict in the leakage field due to surface cracks and distant capillaries. The accuracy and practical application of the MFL system have been experimentally evaluated for detecting surface and remote cracks of a sample made of low carbon steel, with a thickness of 10 mm [9]. A large current around the pipe circumference was supplied, using the Helmholtz coil magnetization method to generate a strong magnetic field saturated around the wall of the pipe. This will result in high permeability in the magnetic infusion, which leads to allowing the sensors to detect the location and size of the cracks and how dangerous they are [10].

A small device has been studied and developed, to generate a magnetic field leakage with a sensor that acts as a receiver to detect cracks on high carbon steel pipes [11]. The test method for permeability measuring magnetic flux leakage PMFL was applied as a non-destructive (NDT) one, showing a good response to the crack shape. It is also sensitive to detecting deep defects, such as the internal or side cracks of the thick steel pipes. More important, an electromagnet is used, flowing through the single coil, from a multi-track loop, where the loop produces a magnetic field required for detection [12]. Hall sensors were used to measure leaks in magnetic fields in the external pipes. Where that sensors receive the crack signal detection for the purpose of obtaining data and then send to the worker and to store and compress this data and reduce noise [13] (Fig. 5).

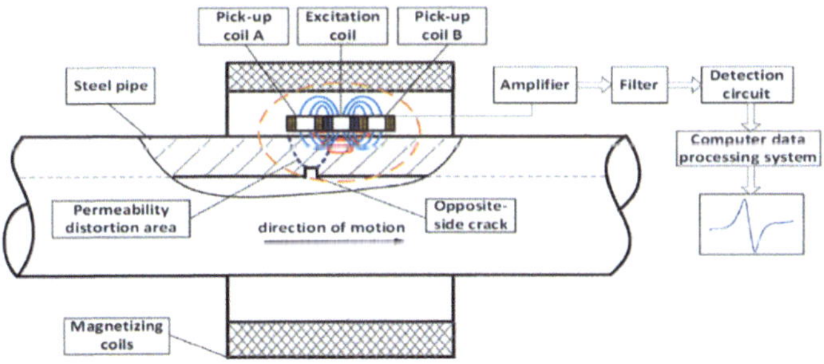

Fig. 5. The schematic diagram of P-MFL testing method [12].

The physical bases of the cracks detection in pipes, when using the magnetic field, depends on its distortion, when the inner surface cracked layer has a low permeability which increases towards the external surface of the pipe, as shown in the Fig. 6, where the pipe usually reaches a magnetic saturation level. The new P-MFL test method was analyzed and demonstrated by the simulations and experiments presented with good results. This method of inspection is similar to the eddy current test, but according to the principle of detection, it is quite different in detail because of the skin effect and the presence of insulating material around the pipes. The detection limit in this P-MFL test exceeds 3 times the depth of the skin in the previous traditional way. However, for the P-MFL test method, the main difference is the fluctuations of permeability in the surface layer caused by deformation of the internal magnetic field lines. The proposed P-MFL test method differs from the previous MFL one from two very important points of view.

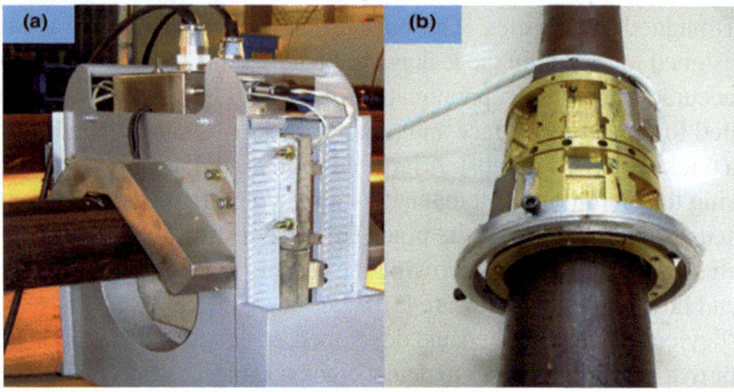

Fig. 6. Practical application of P-MFL testing method [12].

Firstly, although both benefit from the turbulent magnetic field caused by interruptions, the traditional methods measure the initial changes in the magnetic field, while the P-MFL test measures changes in the induced magnetic field, after interaction with the surface layer. The MFL probe senses the physical quantities of the outer space, while the P-MFL test senses directly the property changes within the object. In addition, magnetization and captured signals are separated into different components and are perpendicular to space, but in the MFL test, the captured signals are part of the magnetic dimensional field. Therefore, the P-MFL test method has a higher sensitivity and a large detection range, compared to the conventional MFL test method. To illustrate the advantage of the P-MFL method, a crack with depth 1 mm, and 0.5 mm width was taken in a 14.5 mm thickness sample [12]. The Fig. 7 shows a typical component of magnetic flux density at a lift-off of 1 mm. It can be seen that the amplitude value of B_{MFL} is only 2 mT, while the background magnetic density reaches nearly 225 mT, but the change in small magnetic density cannot be detected by ordinary magnetic sensors. Moreover, it is possible to immerse them in background noise, not to mention complex test conditions in practice. Figure 7 also shows the distribution of relative permeability above the crack. This strong deformation of the permeability of the surface layer makes it easy to make a change in the distribution of the current and thus cause significant induced signals.

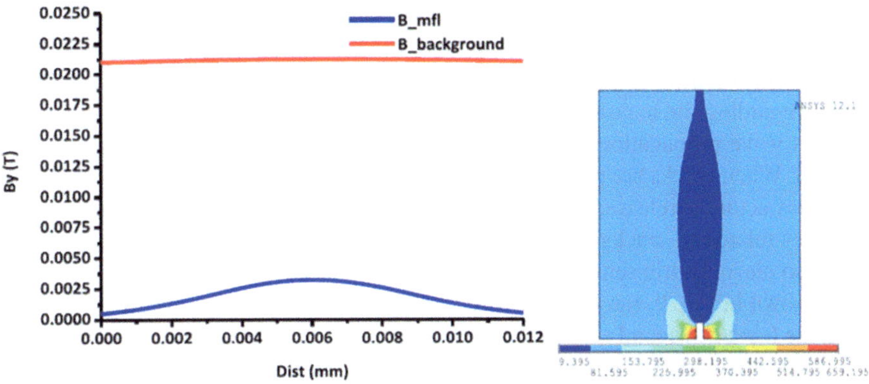

Fig. 7. The magnetic leakage flux density B_y (left) and relative permeability distribution, μ_r in P-MFL [12].

Secondly, because of the principle of interaction, magnetic pressure occurs while magnetic diffusion is disabled in a spatial area filled with a certain level of magnetic flow. Strong field input results in weak MFL outputs, resulting from the magnetic pressure effect (MCE) of the current MFL test sensor. After that, an experiment was performed on the opposite crack system with a depth of 0.5 mm and 1 mm on the opposite side. The test results showed that the P-MFL method was a good response to a 15 mm thickness pipe sample, while there were no clear signals by the conventional MFL method, under normal test conditions.

4 Acoustic Detection of Cracks and Corrosion

Probably, the newest method for detecting a crack with acoustic means was reported in [14]. The experiment was made on a pipe with a drilled hole, as artificial crack, to which an acoustic non-contact excitation was applied by help of laser-induced plasma (LIP) in air. This is an effective way to produce intense sound, without damage of the tested structure. Can realize non-contact acoustic impulse excitations (Fig. 8).

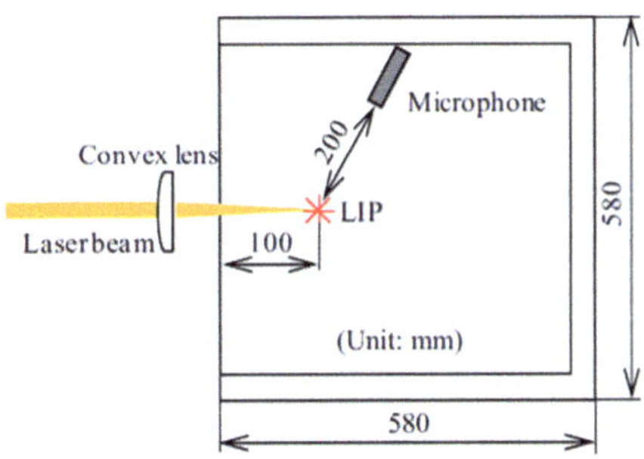

Fig. 8. Acoustic excitation via LIP [14]

A microphone is used to measure the time response of the acoustic pressure, which differs if there is a crack/hole or not in the structure, due to the reflection of the acoustic wave. The position of the crack/hole is detected by applying the wavelet transform to the data regarding the measured time responses in the case of existing hole and not.

Stress wave propagation in solids (sounds and ultrasounds) is also used for crack detection. When cracks are produced and propagate, the high potential energy of deformation or fracture is released with acoustic emissions (AE). An analysis of the detected AE waves related to cracks showed that they have frequencies in the range 50 kHz–2 MHz, so piezoelectric sensors are required [15].

As shown in Fig. 9, the flaw position can be determined if the time delay between hitting the first and second sensor, ΔT is measured and the propagation speed is known for the pipe material (V). The distance between crack and fist hit sensor is:

$$d = \frac{1}{2}(D - \Delta T.V) \quad (3)$$

where D is the distance between sensors.

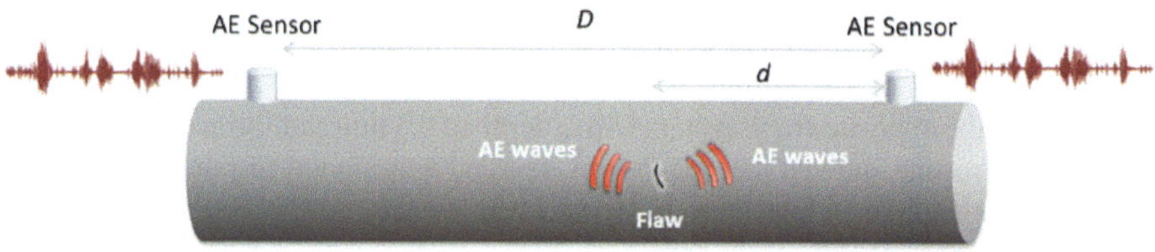

Fig. 9. Detection of acoustic emission in a pipe [15].

It is proved that the propagation of waves due to cracks in the gas pipelines also generate friction, which gives an important signal for the cracks detection. These results provide evidence that sound signals such as wave propagation and Hilbert Huang transform HHT are the best combination to detect cracks [16]. The analysis of acoustic technology is used to detect cracks, but also all obstructions by 10% of the area of the tube and the remote and continuous monitoring of oil and gas pipelines with large diameters [17].

5 Conclusions

By reviewing the most important literature references for three common and important methods for detecting cracks in pipes, relevant advancement was made for the future development of a robotic system for inspection and monitoring of the oil and gas pipes. This is a challenge indeed, if considering the extreme possible environmental conditions as high temperatures in the Middle East countries, by taking into account that first author is from the Republic of Iraq. Therefore, the first type of detection, by use of a thermal camera is the best choice for flaw detection because it's easy to operate, has remote detection capabilities and good performance and reliability.

References

1. Hwang, J., Kim, J., Lee, J.: Magnetic images of surface crack on heated specimen using an area-type magnetic camera with high spatial resolution. In: Instrumentation and Measurement Technology Conference, I2MTC'09, IEEE, 5–7 May 2009, pp. 1546–1551. IEEE (2009)
2. Shafeek, H.I., Gadelmawla, E.S., Abdel-Shafy, A.A., Elewa, I.M.: Assessment of welding defects for gas pipeline radiographs using computer vision. NDT & E Int. **37**(4), 291–299 (2004). (ISSN 0963-8695)
3. Genest, M., Dudzinski, D. C., Bulmer, S., Kersey, R.K.: Crack detection using induction thermography for thermomechanical fatigue tests. In: AIP Conference Proceedings, vol. 1335, pp. 1727–1734. https://doi.org/10.1063/1.3592137
4. Wagner, D., Ranc, N., Batthias, C., Paris, P.C.: Fatigue crack initiation detection by an infrared thermography method. Fatigue Fract. Eng. Mater. Struct. **33**(1), 12–21 (2010)
5. Kostson, E., Weekes, V., Almond, D.P., Wilson, V., Tian, G.Y.: Crack detection using pulsing eddy current stimulated thermography. In: AIP Conference Proceedings, vol. 1335, pp. 415–422. https://doi.org/10.1063/1.3591882
6. Maldague, X., Galmiche, F., Ziadi, A.: Advances in pulsed phase thermography. Infrared Phys. Technol. **43**(3–5), 175–181 (2002). (ISSN 1350-4495)
7. Maffren, T., Juncar, P., Lepoutre, F., Deban, G.: Crack detection in high-pressure turbine blades with flying spot active thermography in the SWIR range. In: AIP Conference Proceedings, vol. 1430, pp. 515–522. https://doi.org/10.1063/1.4716270
8. Ghidoni, S., Minella, M., Nanni, L., Ferrari, C., Moro, M., Pagello, E., Menegatti, E.: Automatic crack detection in thermal images for metal parts. Department of Information Engineering, University of Padova, Italy
9. Okolo, C.K., Meydana, T.: Pulsed magnetic flux leakage method for hairline crack detection and characterization. Wolfson Centre for Magnetics, School of Engineering, Cardiff University, UK (2017)
10. Wu, J., Fang, H., Li, L., Wang, J., Huang, X., Kang, Y., Sun, Y., Tang, C.: A lift-off-tolerant magnetic flux leakage testing method for drill pipes at Wellhead. Electronic Engineering, Newcastle University, UK (2017)
11. Ibrahim, M., Zakaria, Z., Aiman, A.M., Desmond, D.S., Nasir, M.A., Balkhis, I., Ruzairi, R.R.: Magnetic flux leakage capsule material selection and its robustness analysis in oil and gas pipeline. Sens. Transducer **156**(9), 67–74 (2013)
12. Deng, D., Sun, Y., Kang, Y., Song, K., Wang, R.: A Permeability-Measuring Magnetic Flux Leakage Method for Inner Surface Crack in Thick-Walled Steel Pipe. Springer Science (2017)
13. Shi, Y., Zhang, C., Li, R., Cai, M., Jia, G.: Theory and application of magnetic flux leakage pipeline detection. Sensors **15**, 31036–31055 (2015)
14. Kajiwara, I., Akita, R., Hosoya, N.: Damage detection in pipes based on acoustic excitations using laser-induced plasma. Mech. Syst. Signal Process. **111**, 570–579 (2018)
15. Integrity Diagnostics homepage: http://www.idinspections.com. Last accessed 2018/06/02
16. Adnan, N.F., et al.: Leak detection in the gas pipeline by acoustic and signal processing - a review. In: IOP Conference Series: Materials Science and Engineering (2015)
17. Tezerjani, A.D.: High-resolution visual pipe characterization system using an omni-directional camera. Ph.D., thesis, University of Regina, Saskatchewan (2015)
18. Jacquey, F., Comby, F., Strauss, O.: Non-additive approach for omnidirectional image gradient estimation. France (2007)

Design and Analysis for Motors Control of a 4-DOF Parallel Robot

Tudor Catalin Apostolescu[1], Georgeta Ionascu[2(✉)], Silviu Petrache[2], Lucian Bogatu[2], and Laurentiu Adrian Cartal[2]

[1] Faculty of Informatics, Titu Maiorescu University, Bucharest, Romania
[2] Faculty of Mechanical Engineering and Mechatronics, POLITEHNICA University of Bucharest, Bucharest, Romania
ionascu_georgeta@yahoo.com

Abstract. In this paper, the complex robotic structure of a new 4-DOF parallel robot is designed in a mechatronic concept that integrates mechanical and control aspects. The kinematic and dynamic analysis for the control of motors that act the robot's arms are presented. Delphi, LabView and SolidWorks softwares are used. The adopted constructive solution of the robot, designed in SolidWorks, has been presented. Then, the transposition of equations which describe the forward and inverse kinematic model of the robot has been performed using the LabView programming environment, a program that allows calling the servomotors' control board - NI 7344 acquisition board. The angles for the actuated arms of robot have been calculated based on a program developed in the Delphi environment that contains both models - the forward and the inverse one. In order to determine the motor load, a dynamic computation and a simulation have been achieved in ComosMotion from the SolidWorks programming environment. Different results were extracted in the form of diagrams: the torque for the three servomotors acting the arms, the most loaded servomotor – the one that drives arm 1 and the variation of the angular velocity for the arm 1. Finally, the robot prototype has been achieved and interesting conclusions regarding the analyzed robotic structures have been inferred.

Keywords: Mechatronic design · Parallel robot · Motion control

1 Introduction

At the moment, most industrial robots are made with tree structure, in this case each axis is positioned in series to the previous one, and therefore we can say that the axes are positioned in line to each other. The advantage of such a structure is its simplicity, being easily fabricated and controlled. The disadvantage of these structures is that unlike the parallel structures, are less rigid.

In case of parallel robots, the actuators are arranged parallel to each other and aim to orientation and positioning of a mobile platform on which there is placed the end-effector. The drive for this type of robot can be linear or rotary. The advantages of such a structure are: stiffness, it is a more rigid structure than in those in the series, which is

G. I. Gheorghe (Ed.): ICOMECYME 2018, LNNS 48, pp. 194–206, 2019.
https://doi.org/10.1007/978-3-319-96358-7_20

due to closed kinematic structure; kinematics increasing, high speeds and accelerations. The disadvantage of such a structure is its complexity.

Based on the Stewart-Gough platform architecture, a significant number of mechanical devices have been proposed [1–3]. In many industrial applications, the systems with 4 DOFs ("Degrees of Freedom") are sufficient: 3 translations to move the carried object from one point to the other and 1 rotation to adjust the orientation in its final location. Several research performed in this domain, particularly those by Clavel [4], have led to innovative architectures such as the famous Delta robot [5, 6].

In order to develop such a complex robotic structure, a mechatronic concept that integrates mechanical and control aspects has been considered. The design of a new 4-DOF parallel robot and, also, the kinematic and dynamic analysis for the control of motors that act the robot's arms are presented in this paper.

2 Presentation of the Adopted Constructive Solution

Usually a parallel 4-DOF robot is achieved by using a linear or rotary actuating system, a fixed part which forms the base of the robot, three parallelogram mechanisms for the parallel movement of a mobile plate or of a mobile platform that has the role of positioning the effector and the working object (WO).

It also needs an orientation system that can be achieved by using an actuation made by means of a servomotor which can be placed on the fixed plate or on the mobile one and an effector.

Another problem in designing this type of parallel structure is the location of the effector action, or in other words, the location of the fourth degrees of freedom.

In order to design the robot there were chosen spherical joints, rotary motors using harmonic gearboxes and the end effector was posed on the robot mobile platform.

For the robot designing the SolidWorks software has been used. The figures from below are captures representing the final assembly of the parallel robot structure with 4-DOF.

The components of the rigid or fixed robotic structure (1–5) and of the mobile subassembly (6–13) are shown in Fig. 1, as follows: 1 – frame made from square bars; 2 – gusset plate; 3 – corner; 4 – the robot base or fixed upper plate; 5 – fixing rods, which make the connection between the frame and the robot base; 6 – gear motors; 7 – robot arm; 8 – passive joints, the spherical coupling model used being an open one in a constructive version called "bar head" (detailed in Fig. 2); 9 and 10 – bars or rods that form a parallelogram; 11 – the mobile platform or lower plate; 12 – the effector; in the proposed solution it consists of a suction cup; 13 – WO.

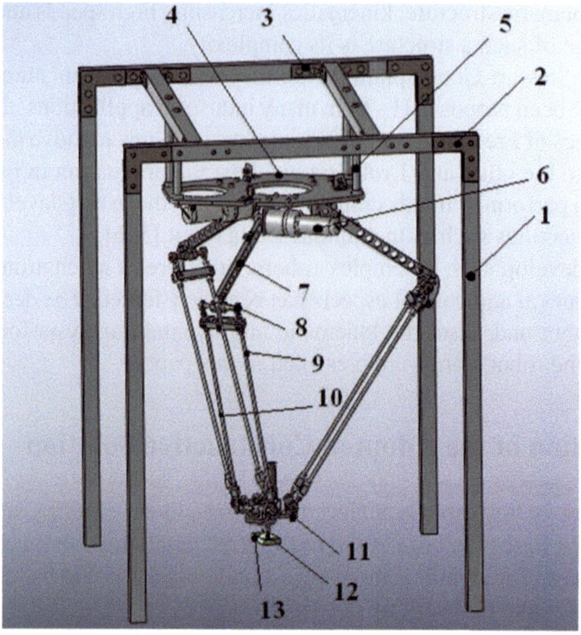

Fig. 1. Assembly of the new 4-DOF parallel robot, represented in SolidWorks.

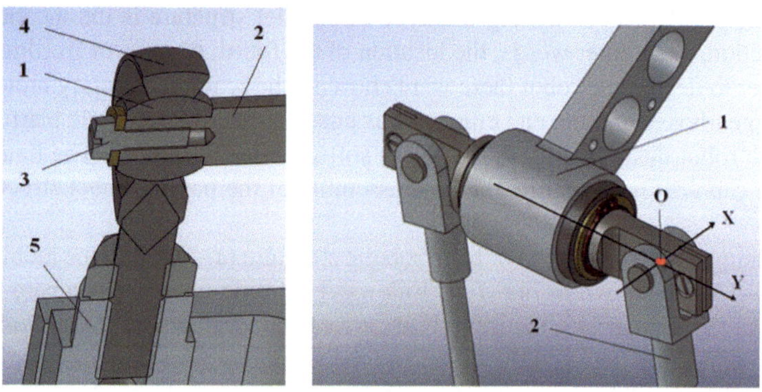

(a) (b)

Fig. 2. Type of joints: (a) the spherical joint using 1 – inner sphere, 2 – spindle, 3 – assembling screw, 4 – outer sphere, 5 – threaded bushing; (b) the cylindrical joint with 1 – arm, 2 – rods.

To achieve the kinematics of the parallel robot with four degrees of freedom, it is necessary to use three similar subassemblies, each one consisting of a gear motor, arm, spherical joints and rods forming a parallelogram mechanism.

The spherical joints can be replaced with two cylindrical joints with perpendicular axes. Figure 2 shows the construction variant for the replacement of spherical joints with 2 cylindrical joints with perpendicular axes.

The spherical joint (Fig. 2a) between the two elements 1-arm and 2-rods with the center in O is replaced by two cylindrical couplings with OX and OY axes (Fig. 2b).

To achieve the parallel robot with four degrees of freedom, we have chosen the spherical joint with its open version because of the precision and mobility considerations.

Figure 3 allows a better understanding of such a subassembly: 1 – the robot base or fixed upper plate; 2 – motor corner; 3 – gear motor; 4 – embedded screws for fixing the gear motor; 5 – robot arm; 6 – clamping screws of the arm on the gear motor shaft; 7 – outer sphere of the spherical joint; 8 – corner of microswitch used to limit the stroke; 9 - end stroke (called "Forward") microswitch that blocks dangerous, unpredicted movement of the mobile platform; 10 – part used for preventing the rotation and, at the same time, for actuating the microswitch if it is necessary.

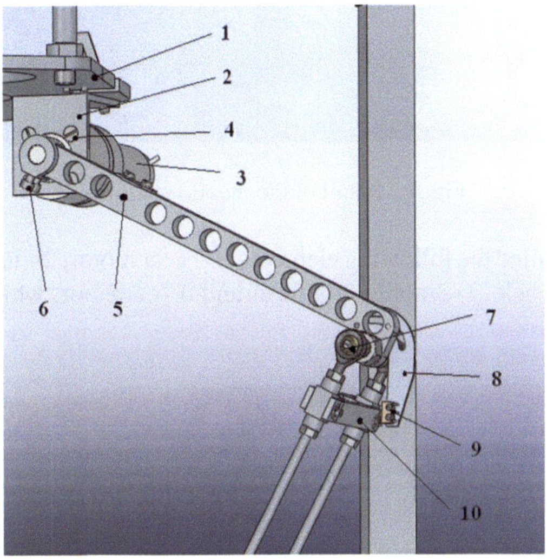

Fig. 3. Detail of the robot arm.

Figure 4 shows a detail of the mobile platform and its component elements: 1 – the lower plate or mobile platform; 2 – orientation box; 3 – wheel belt mounted on the output shaft of the gear motor; 4 – toothed belt; 5 – shaft - belt wheel; 6 – ball bearing; 7 – adapter for the suction cup; 8 – the effector represented by a suction cup; 9 – the working object (WO).

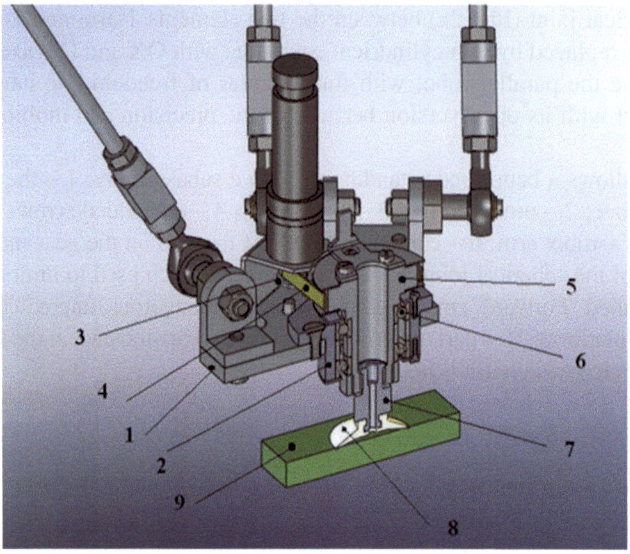

Fig. 4. Detail of the mobile platform.

In Fig. 5 are detailed the following elements: 1 – gear motor, 2 – toothed wheel driven by 5 – the toothed belt, 3 – mobile platform and 4 – effector, which in this case is a suction cup.

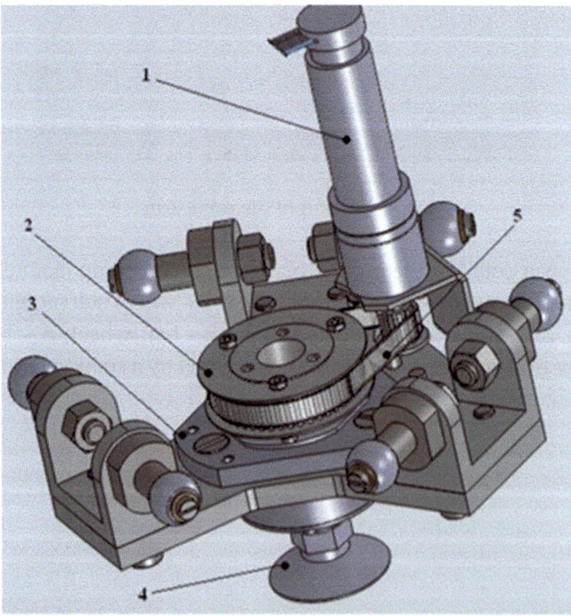

Fig. 5. Sub-assembly represented by positioning of the effector actuation on the mobile platform

3 The Computation Memoir

3.1 The Computation of the Forward and Inverse Geometric Model

The Delphi program does not have the ability to work with the NI 7344 servomotors control board. Therefore, was needed a similar program to that created in Delphi but written in a compatible programming environment.

The transposition of equations which describe the geometric forward and inverse model [7] was performed using the LabView programming environment, a program that allows calling the servomotors control board - the NI 7344 acquisition board.

The purpose of this program is to be called as a subfunction by LabView programs designed to control the robot. Figure 6 shows a simple way to call the sub-function. There are used three controllers for introducing the Cartesian coordinates x_0, y_0 and z_0 and also are employed three indicators for displaying the values of the resulting angles θ_1, θ_2 and θ_3.

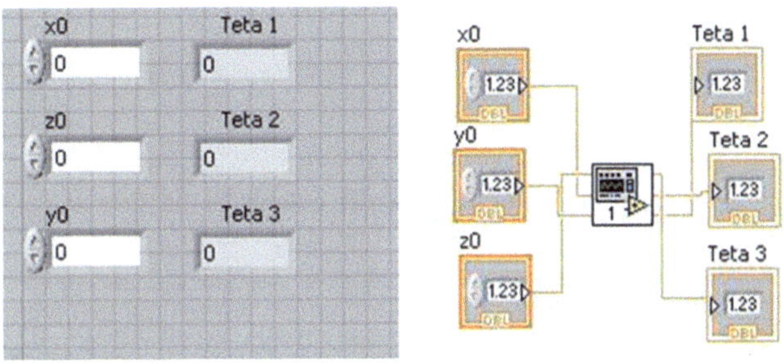

Fig. 6. The call of the subvi "sub vi program.vi". On the left is the control panel and on the right is the block diagram of the program.

In a complex program, instead of the three controllers, are linked the values obtained by that program for the three Cartesian coordinates. Those three outputs are linked to the motion vector generation subprogram, and the resulting motion vector is sent to the servomotors motion control board.

Below is given a part of the computation program, which calculates the theta angle 3. In the block diagram, the output variable is called "Delta 3". Because the block diagram could not be included as a single figure, it was decomposed into 3 figures: Figs. 7, 8 and 9.

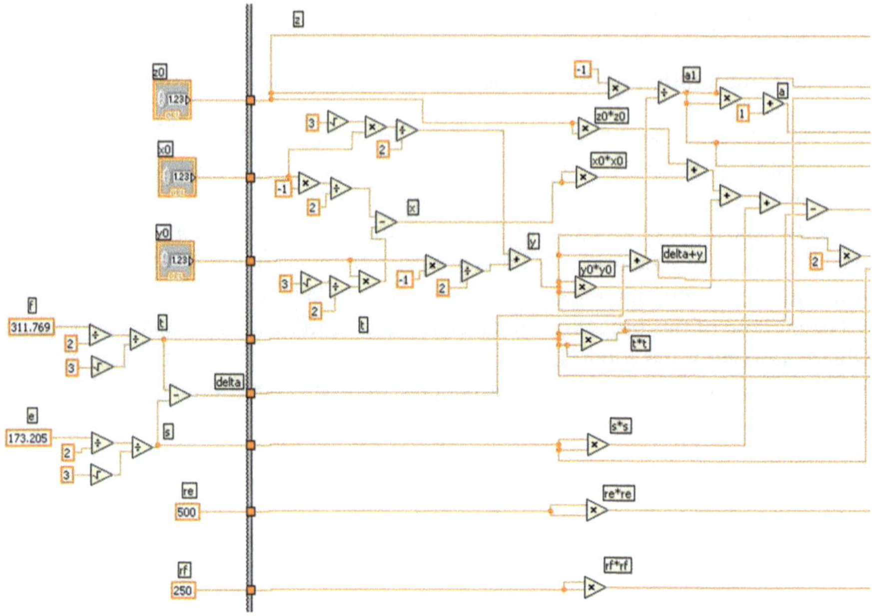

Fig. 7. The block diagram for the subvi which calculates the theta angle 3 - part 1.

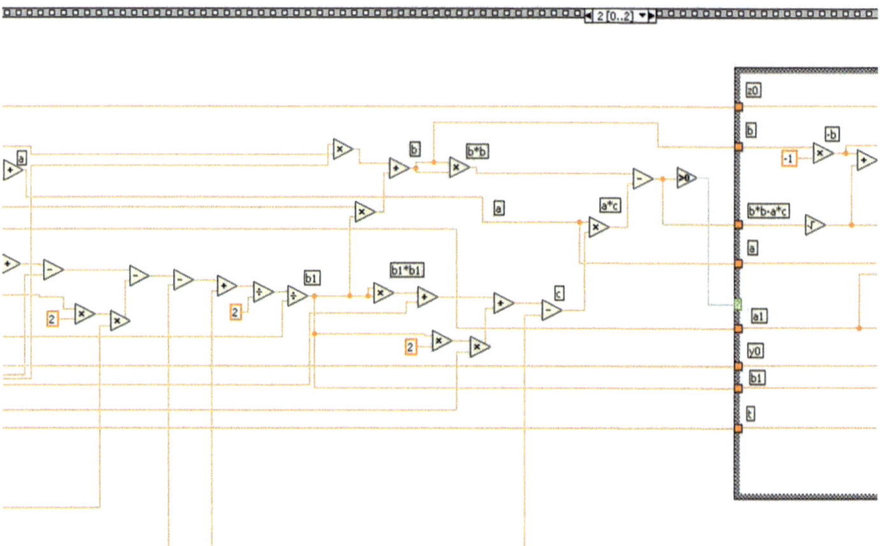

Fig. 8. The block diagram for the subvi which calculates the theta angle 3 - part 2.

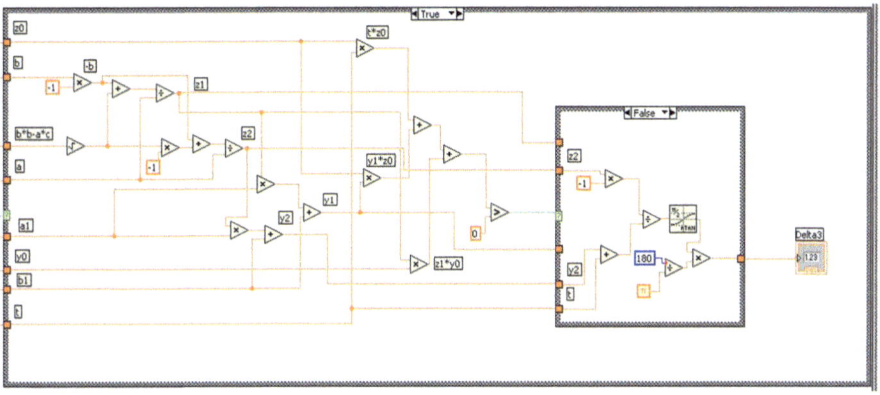

Fig. 9. The block diagram for the subvi which calculates the theta angle 3 - part 3.

3.2 The Computation of the Robot Arms' Angles

In order to calculation the arms' angles, we assume a linear trajectory between two points, initial I(x_I, y_I, z_I) and final F(x_F, y_F, z_F) with the following coordinates (the position of the axes is according to Fig. 10): $x_I = -25$ mm; $y_I = -25$ mm; $z_I = -550$ mm; $x_F = 25$ mm; $y_F = 25$ mm; $z_F = -450$ mm.

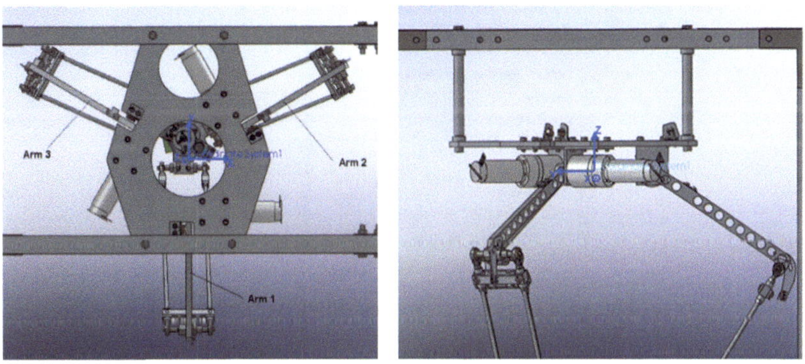

Fig. 10. The position of the axes.

The calculation of the θ angles for the actuated arms is based on the inverse geometric model. A "ProjectDelta2" program is developed in the Delphi environment that contains both models - the forward and the inverse one.

The values of the angles θ_1, θ_2 and θ_3 are given in Table 1. In the table are also mentioned the Δθ variations of the angles related to the values corresponding to the initial point I.

Table 1. The values for the angles θ_1, θ_2 and θ_3.

	Units	Point I (initial)	Point F (final)
x	mm	−25	25
y	mm	−25	25
z	mm	−550	−450
θ_1	deg	−26.5	−13.2
θ_2	deg	−33.4	−4.6
θ_3	deg	−28.4	−10.9
$\Delta\theta_1$	deg	0	13.3
$\Delta\theta_2$	deg	0	28.8
$\Delta\theta_3$	deg	0	17.5

3.3 Simulation of the Arms' Behavior

To determine the engine load a dynamic computation and a simulation were performed in ComosMotion from the SolidWorks programming environment.

In the simulation are inserted the servomotors (RotaryMotor) that actuate the arms - Fig. 11.

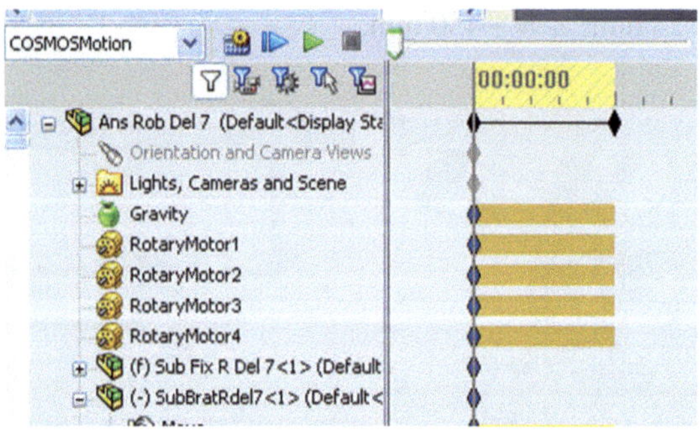

Fig. 11. The CosmosMotion simulation settings.

In Fig. 12 the MR1 servomotor (in RotaryMotor1 simulation) is shown to actuate the arm 1 of the structure.

The motion kinematics is of the "step" type without acceleration leaps (Fig. 13). The angular acceleration curve ε has no discontinuities and its variation is linear. Growth occurs during Δt. The angular velocity curve has parabolic variation, both on the acceleration zone and during the braking period.

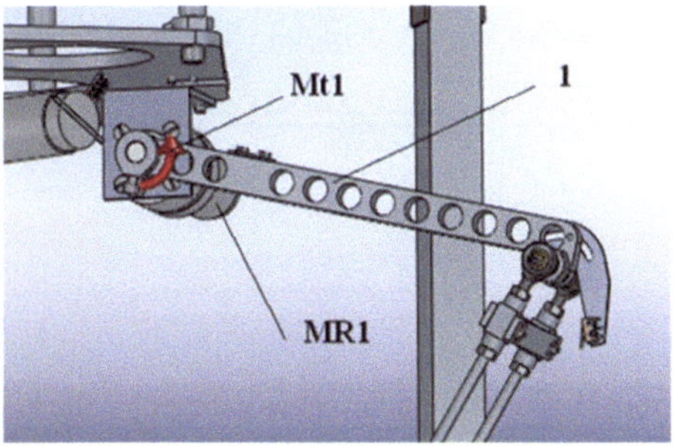

Fig. 12. The CosmosMotion simulation – the MR1 servomotor actuating the arm 1.

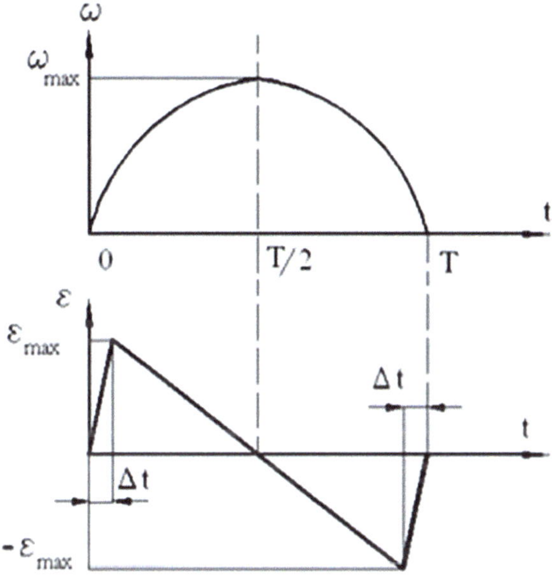

Fig. 13. The CosmosMotion simulation – the angular velocity and acceleration.

The settings of this motion for the simulation is shown in Fig. 14, where there are indicated the all the data for the kinematic of the RotaryMotor1 (MR1). The step function includes the arguments: time, initial time (0 s), initial angle (0D), final time (0.5 s) and final angle (13.3D).

After completing the simulation, different results can be extracted in the form of diagrams. Figures 15, 16 and 17 show the torque for the three servomotors acting on the arms: RotaryMotor1, RotaryMotor2 and RotaryMotor3.

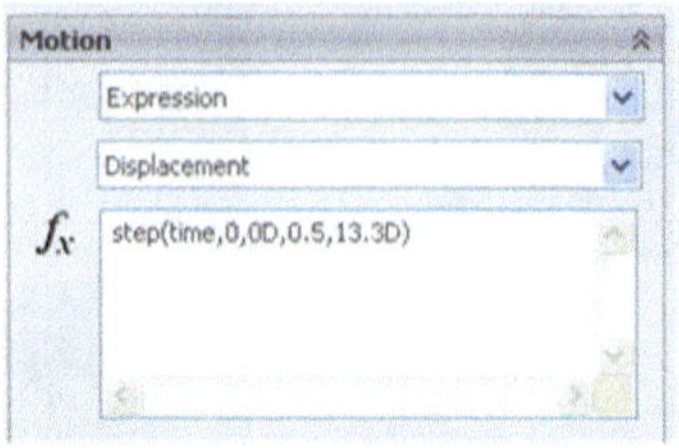

Fig. 14. The CosmosMotion simulation – the settings for the kinematic of MR1.

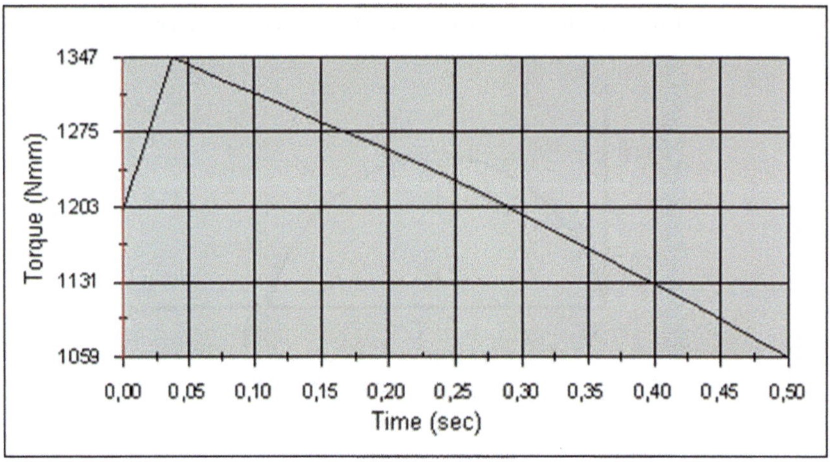

Fig. 15. The CosmosMotion simulation – the torque for MR1.

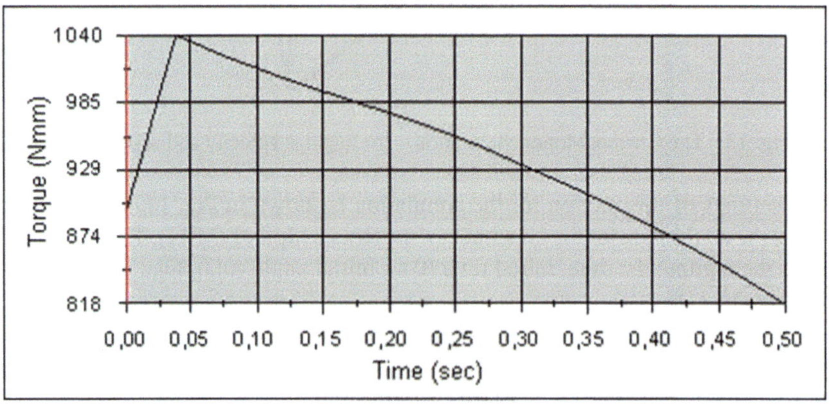

Fig. 16. The CosmosMotion simulation – the torque for MR2.

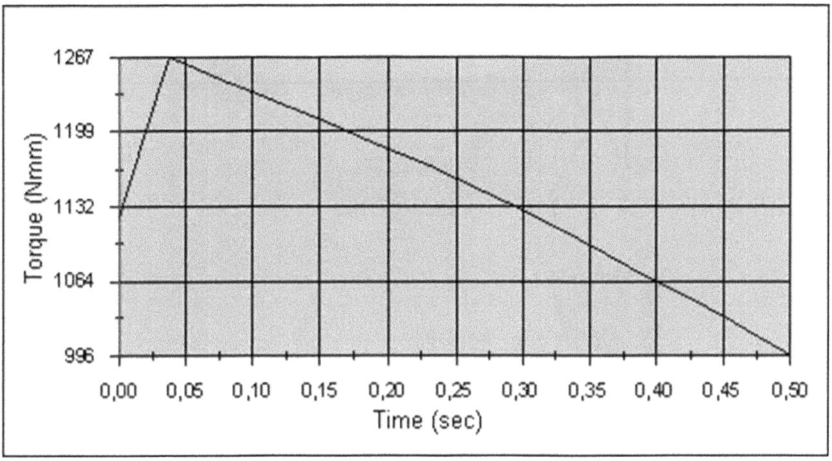

Fig. 17. The CosmosMotion simulation – the torque for MR3.

From chart analysis, the most loaded servomotor is the one that drives arm 1 (RotaryMotor1) requiring torque Mt = 1347 Nmm.

Figure 18 shows the variation of the angular velocity of the arm 1.

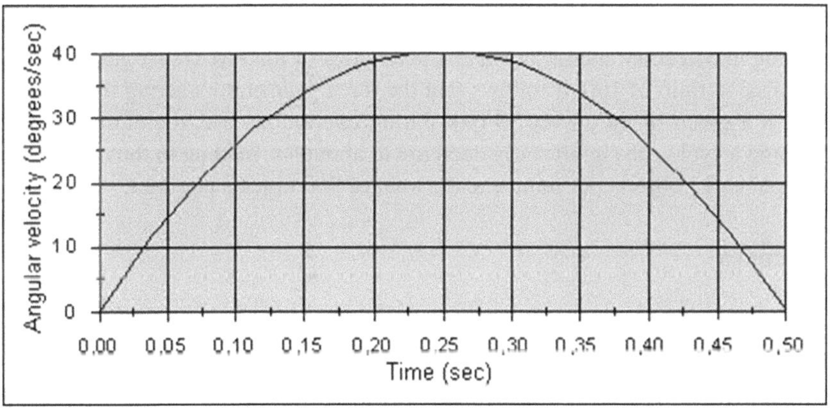

Fig. 18. The CosmosMotion simulation – the angular velocity of the arm 1.

4 Conclusions

The feasibility of the original designing solutions for the new 4-DOF parallel robot has been demonstrated through achieving the prototype of this robot, Fig. 19. Two kinds of experiments have been performed: manipulating the objects and image analysis. The robot is able to detect new objects in the workspace and it can achieve object orientation even if this requires a rotation of more than 90°.

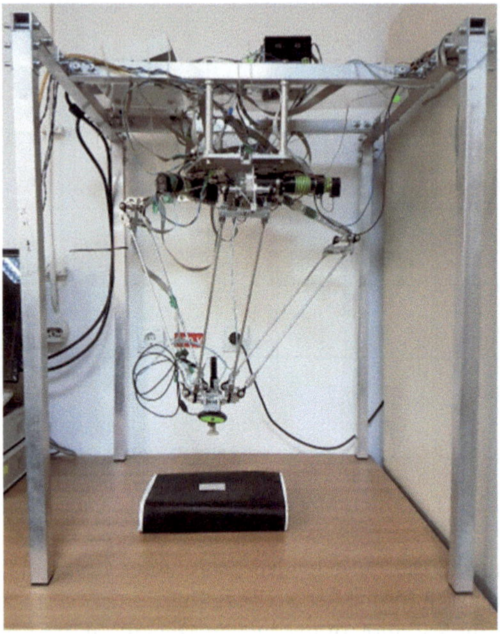

Fig. 19. Prototype of the new 4-DOF parallel robot.

Taking into account the technical characteristics of the PMA-14A gear motor with a transmission ratio of 100, it follows that the three gear motors acting the robot arms can reach a speed of little over 30 rpm during continuous use, therefore the time of performing a cycle can significantly decrease to about 4 s, leading to the robot's ability to manipulate 15 objects per minute and a total of 900 objects per hour.

References

1. Udrea, C., Alexandrescu, N., Panaitopol, H., Avram, M., Apostolescu, T.C.: The constructive bases of industrial robots (in Romanian). Editura Universitara Publishing House, Bucharest (2006)
2. Merlet, J.P.: Parallel Robots, 2nd edn. Springer, Dordrecht (2006)
3. Siciliano, B., Khatib, O. (eds.): Handbook of Robotics, 2nd edn. Springer, Berlin (2016)
4. Clavel, R.: Device for the movement and positioning of an element in space. U.S. Patent 4976582 (1990)
5. Poppeová, V., Uríček, J., Bulej, V., Šindler, P.: Delta robots – robots for high speed manipulation. Techn. Gaz. **18**(3), 435–445 (2011)
6. Stan, S.D., Manic, M., Szep, C., Balan, R.: Performance analysis of 3 DOF delta parallel robot. In: HSI 2011 (4th International Conference of Human System Interaction), May 19–21, Keio University, Yokohama, Japan (2011)
7. http://forums.trossenrobotics.com/tutorials/introduction-129/delta-robot-kinematics-3276/. Accessed 30 May 2016

Intelligent Mechatronics and Cyber-Mechatronics Ecosystems Developed in "ECOSIN - MECATRON" Research Infrastructure

Gheorghe I. Gheorghe(✉)

The National Institute of Research and Development in Mechatronics and Measurement Technique, Pantelimon no. 6-8, sector 2, Bucharest, Romania
geocefin@yahoo.com

Abstract. Within the framework of the "ECOSIN – MECATRON" Research Infrastructure project, INCDMTM - Bucharest develops Intelligent Mechatronics and Cyber - MixMechatronics Intelligent EcoSystems, which is the foundation for the Intelligent Mechatronics Industries for the Intelligent Field of Specialization and for Mechatronics and Cyber - Mechatronics, by presenting advanced solutions and intelligent technologies related to the economic, industrial and social sustainability in Romania.

Keywords: Intelligent ecosystems
Mechatronics & Cyber-Mechatronics Systems · Research infrastructure
ECOSIN-MECATRON

1 Motivation

This scientific paper is motivated by "the scientific paradigm of Mechatronics, Mix – Mechatronics and Cyber – MixMechatronics", meaningful and descriptive thesaurus integrated into the architecture of global systems [1].

Thus, the structure and new intelligent specialization of Mechatronics, Mix – Mechatronics and Cyber – Mechatronics, treats the philosophy of education in the new Engineering for the future and future cognitive society, as well as the holistic multidisciplinarity of it.

Depending of the new requirements of the process, product, materials, technology, energy, application an effect in industry, economy and society, this new multidisciplinary domain – Mechatronics has been developed, upgraded and adapted to meet different operational criteria, such as the reduction of material, energy and financial resources, high-level operation, optimal ergonomics and cost-cutting at the limit.

In this sense, the concept and the evolutionary structure of the intelligent Mechatronics field develops towards a generative evolution by adding new structural and functional elements, components and subsystems on new principles and adapted to new scientific discoveries in order to obtain technical facilities and effects, technological and economic demands required by the concrete applications of adaptive mechatronic products, technologies and services.

G. I. Gheorghe (Ed.): ICOMECYME 2018, LNNS 48, pp. 207–223, 2019.
https://doi.org/10.1007/978-3-319-96358-7_21

Thus, in Fig. 1, are presented the concept and product/mechatronic adaptronic system.

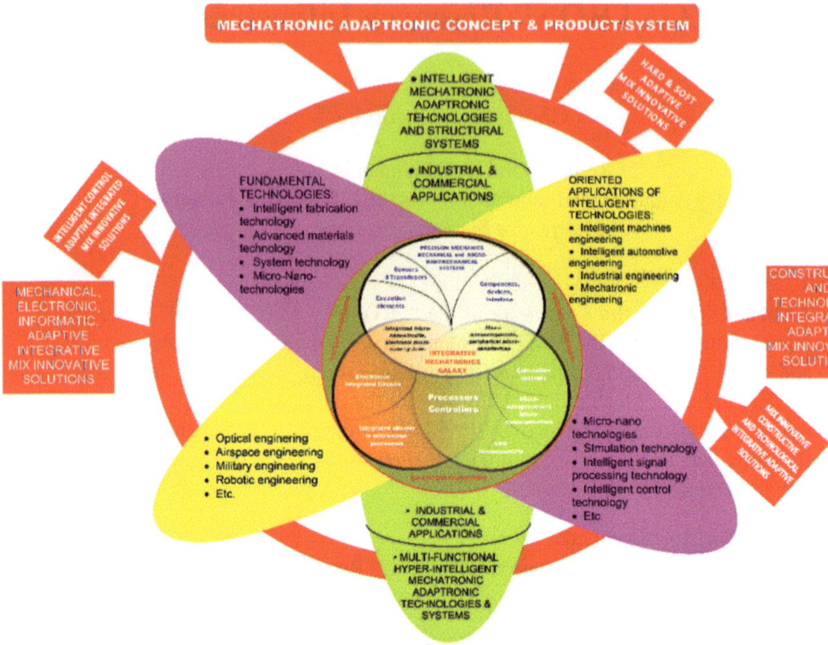

Fig. 1. The concept and product/mechatronic adaptronic system

According to Fig. 1, the concept and product/mechatronic adaptronic system are synthesized by the synergy of mechanical subdomains, including μmechanics and nmechanics (systems, devices, mechanical links, components, etc.), electronics, including electronics and non-electronics (components, circuits hard drives, computing systems, processors, controllers, software systems, etc.) and information technology, including microinformatics and nanoinformatics (computer systems, simple and integrated, peripheral devices, interfaces, modules, sensors, transducers, amplifiers, transformers); the adaptive process for structural and functional development integrates new requirements, conditions and effects to meet new demands, innovative mechanical, electronic and integrated integration adaptive mix solutions, innovative integrated intelligent integrated adaptive control solutions and innovative hardware and adaptive software solutions; adaptation process, integrates the fundamental technologies for developing high-performance technology, such as intelligent manufacturing technology, advanced materials technology, system technology, micro-nanotechnologies, advanced simulation and modeling technology, intelligent signal processing technology, intelligent control technology, etc.; intelligent technology applications include a wide range of scientific fields, such as smart machine engineering, intelligent car engineering, industrial engineering, mechatronics engineering, optical engineering, aerospace engineering, military engineering, robotics engineering, etc.; applications of adaptive products/systems (mechatronic) are synthesized in their entirety in industrial applications and commercial applications; the advanced results of advanced high-tech Adaptronics Mechatronics are materialized in adaptive (mechatronic) technology and systems, and in adaptive and adaptive multi-functional hi-tech (mechatronic) technologies and systems that contribute to the sustainable development of the intelligent industry, the knowledge-based economy and the information society.

The high-tech Mecatronics Adaptronics domain, Mixed - Mecatronics, has been developed step by step, according to the technical and technological requirements of the new generation of intelligent products and technologies and new industrial and commercial applications, assembling innovative vectors from the perspective of mechatronics and mix - mechatronics in a mix-innovative structure and in an integrated systemic conception, cumulating the constructive-functional-decisional relationships similar to the human organism and evolving into a vision of the global-innovative-adaptive concept and product/adaptronic integronic system (Fig. 2), [2, 6, 8, 10].

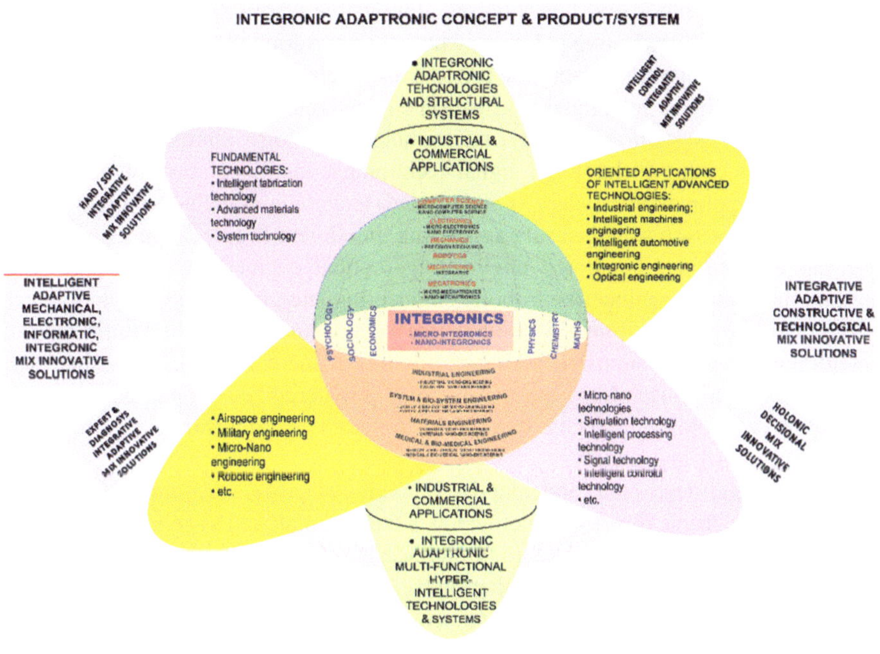

Fig. 2. The vision of the global-innovative-adaptive concept and product/adaptronic integronic system

According to Fig. 2, the advanced intelligent high-tech field is based on the adaptive (integrated) concept/system integrating multi-interdisciplinary industrial and commercial technical and technological solutions with structures, materials and systems intelligent, complete and adaptively assembled.

The resulting new Integronics Adaptronics domain based on Integronics, μIntegronics and nIntegronics is synthesized by the synergy and the highly technologically assembled combination of mechatronic subdomains (including mmechatronics and nmecatronics), integrative mechatronics (including sensory, actuator, digital interfaces, etc.), robotics and mechanics, precision mechanics (including mechanical and precision nmechanics), electronics (including electronics and non-electronics), computer science, and other sciences (physics, chemistry, mathematics, economics, sociology, psychology, etc.), in an industrial-commercial applicative structure with industrial engineering (including industrial engineering), engineering of systems and bio-systems), engineering of intelligent materials (including intelligent materials engineering) and medical and biomedical engineering (including medical nursing and biomedical); the "magistral" process of integration and full adaptation for the creation and development of holonic and active-adaptive system assemblies for new demands, conditions and effects, assembling structural-functional-decision-informational new innovative constructive and adaptive integrative mix innovative mix solutions - innovative, adaptive intelligent mechanical, electronic, intelligent (Integral), new innovative solutions intelligent integrated adaptive control, innovative adaptive integrated hardware and software solutions, innovative mix solutions - innovative adaptive expert and diagnostics integrated and new innovative solutions - holistic innovation decision-making; At the same time, this integrative adaptation process also includes the fundamental technologies used in intelligent integrated intellectronic manufacturing such as advanced materials technology, system technology, micro-nano technology, simulation technology, intelligent processing technology, signal technology, intelligent control technology, etc.; oriented applications of adaptive (integrated) technologies include a broad array of applicative fields such as industrial engineering, intelligent engine engineering, intelligent car engineering, integrated engineering, optical engineering, aerospace engineering, micro-nano engineering, robotic engineering, etc.; Integrated adaptation technology and Integral Intelligent Structural Intelligence and Intelligent Multifunctional (Integronic) Integral Systems are all industrial and commercial applications; the results of the advanced high-tech Integronica Adaptive are materialized in Intelligent Systems, Intelligent Technologies, which contribute to the intelligent development of the future industry, economy and society.

The new, high-tech smart and adaptive domain, Adaptronics/Applied Adaptronics, as a Mix-Mechatronic domain, is defined as a combination and synergy of innovative and multidisciplinary multi-technology engineering that integrates new fundamental knowledge from Multistructural Mechatronics, Multifunctional Integronics, Advanced Intelligent Materials Science, Sensor and actuator architecture, Informatics, Intelligent Measurement Technology and Automatic Intelligent Control (Fig. 3), [3, 7, 9].

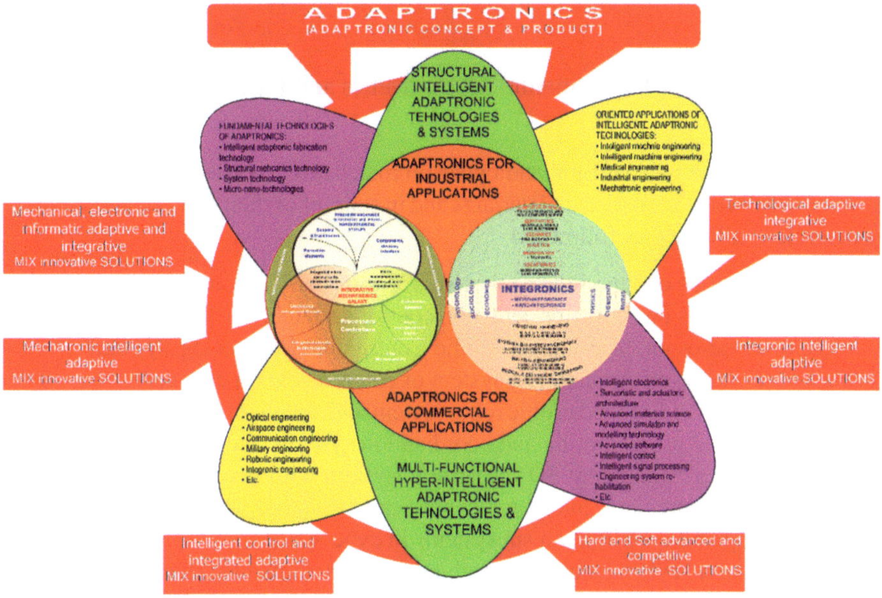

Fig. 3. The new and advanced high-tech field of Adaptronics/Applied Adaptronics

According to Fig. 3, the new and advanced high-tech field, Adaptronics/Applied Adaptronics is defined as follows:

- is «technical-technological system integrator between Research-Development-Innovation-Industrial and Commercial Application and the active exchange of experience and knowledge transfer with the latest solutions for products and systems with adaptive components and structures and with current, evolutionary and generative tendencies and the active exchange of experience and knowledge transfer with the latest solutions for products and systems with adaptive components and structures and with current, evolutionary and generative tendencies»;
- is «synergistic and integrative combination of intelligent technical and technological fields known as intelligent systems, intelligent materials, intelligent processes that describe the easy way to build adaptive integrated systems and structures with the objective of reducing material, technological and energy resources, for implementation and operation to an absolute minimum with application scenarios and focused on nature and trying to simulate vital functions, by the ability of biological systems to automatically recognize and correct dysfunctions in their own structure, this feature being desired in the first line, for the systems and technical structures in the essential areas of safety».

2 Realization

2.1 The Concepts of Experimental Mechatronics and Cyber – Mixmechatronics Models

The concepts of the mechatronic and cyber-mixmechatronic experimental models are involved in selective scenarios and variants and in support of their construction, through engineering challenges, as follows [4, 5]:

- selected scenarios on "monitoring system of controllers from distance";
- selected scenarios on "actuator detection control system";
- selected scenarios on "Automated Traffic Control and Monitoring System";
- selected scenarios on "monitoring system of communication channel";
- the concept of a "cyber-mecatronic robot system with telemonitoring and telecontrol" and modularised programming and coordination systems in cyberspace;
- the concept of a "intelligent 3D mecatronic system with two grippers and two 3D sensors for measurement, integrated control and industrial services" and monitored programming and coordination subsystems in cyberspace;
- the concept of a "cyber-mecatronic technological instrument system with telemonitoring and telecontrol and modularized programming and coordinating subsystems in cyberspace;
- the concept of "convergence of complexity in cyber-mechatronics";
- "microcontroller programming and interfacing to intelligent mechatronic and cyber-mixmechronic systems";
- "hardware and software structures for mechatronic and cyber-mechatronic systems";
- "creating virtual prototypes that increase the speed of development of mechatronic and cyber-mechatronic machines";
- the structure of cyber-mixmecatronic 3D multi-application system with telecontrol and telemonitoring";
- etc.

It is presented in the original concept, the foundation of the multi-application cyber-mix-mechatronic systems as parts of the physical-cybernetics systems (cyber-physical and intelligent solutions for construction). The paper focuses on the basics of cyber-physical systems supported by the world's great strategists, with specific examples and concepts, with potential societal applications - mechatronic and clatronic intelligent systems, virtual reality systems with some major challenges on reliability and uncertainty, abstraction levels and mismatch physics-cybernetics as well as design software, time-sensitive programming languages and networking for superdens time.

The scientific work exemplifies the creation/realization of multi-application cyber-mixmechatronic systems as parts of global cyber-physical systems.

The Universe of Cyber-Physical Systems (or Physical Cybernetics) includes, in its program of development and implementation, the main stages, step by step, to materialize and implement them in society, to ensure innovative national, European and international strategies for a postinformed and neural society.

Thus, they are based on the "bases of cyber-physical systems" created and grounded by the great strategists of the world regarding:

- **The concepts** of cyber-physical systems (supported by global strategists Christopher Chadwick, Sarah Betzig şi Fei Hu);
- **The design challenges of cyber-physical systems** (supported by global strategists Cameron Patterson, Roger Vasquez şi Fei Hu);
- **Creating** mobile cyber-physical systems (supported by global strategists Yeqing Wu şi Fei Hu),

Follows the "design principles of cyber-physical systems":

- Cyber-physical systems controllers (supported by Tony Huynh, Ahmed Alsadah and Fei Hu);
- Learning apprenticeship on the Physical-Cybernetic Intelligent System (supported by Kassie mccarley, Joseph Pivscan and Fei Hu);
- **Application** of HDP-HMM for the dynamic recognition of "hand gestures" (supported by world strategies Lv Wu Ting Zhang and Fei Hu);
- **Modeling problems** in Cyber-Physics Systems (supported by global strategies Michael Johnson, Tony Randolph and Fei Hu);
- **Modeling** Cyber-Physics Systems (Cognitive Vehicles - Remote Airplane), (backed by global strategy Meng Cheng Ong, Fei Hu and Yang-Ki Hong);
- **Security** of Cyber-Physics Systems (supported by Steven Guy, Erica Boyle and Fei Hu);
- **Physical Cyber Security Systems** - Smart Network example (supported by Rebecca Landrum, Sarah Pace and Fei Hu),

Continuing with "Intelligent Sensor Architecture - the basis of Cyber-Physics Systems, on:

- **Wireless sensors** and wireless actuators for applications in Cyber-Physics (supported by Kassie mccarley, Joseph Pierson and Fei Hu);
- **Communication detection** (supported by Trenton Bennet, John Har and Fei Hu);
- **Integrated/Deployed Wireless Microsystem Architecture and Security** (supported by Derek Chandler, Jonathan Pittman, Jaber Abu-Qahouq, and Fei Hu);
- **Application** of a learning machine in physical sensing activity monitoring (Wenlong Tang, Ting Zhong and Edward Sezonov)

Finally, it concludes with "Cyber-Physical Civil Applications" for:

- Creating efficient energy (supported by great strategists Preston Arnett, Jan Wolfe and Fei Hu);
- Creating cyber-physical systems for smart grid smart applications (supported by great strategists Matei Rell, Loilim Muirhead and Fei Hu);
- Creating Video in Unmanned Aircraft for Cyber-Physical Systems (supported by great strategists Meng Cheng Ong, Fei Hu, Yang-Ki Hong, Kenneth Rieks, and Jaber Abu-Qahouq);

2.2 Conception and Realization of Experimental Mechatronics and Cyber-Mixmechatronics Models

- **3D cyber-mix-mecatronic System ultraprecise multiaplicative for remote control and remote monitoring** (Fig. 4), [1, 2, 11].

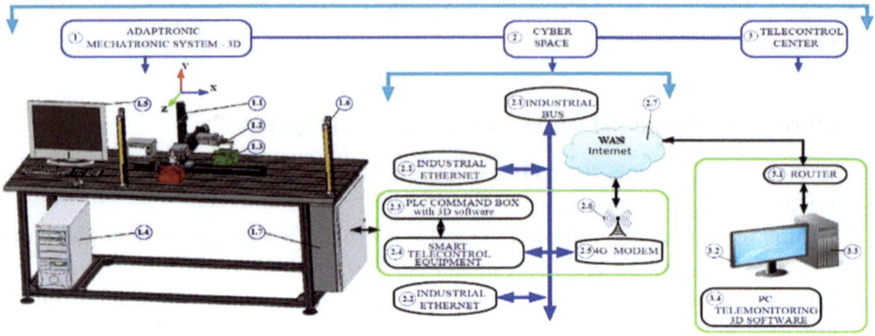

Legend: 1. 3D cyber-mix-mechatronic system:
1.1 Ultra-precise 3D measuring system / measuring robot / ultra precise control robot (x-300mm; y=200mm; z=250mm; accuracy:0.1-1nm);
1.2 3D ultra precise probe (accuracy: 0.1 nm);
1.3 Control / measurement part;
1.4 PC local host;
1.5 Display and local user interface;
1.6 Laser protection barrier
1.7 Unit with command system, driving system and telecommunications system;

2. Auxiliary equipments:
2.1 PLC (Programmable Logic Controller);
2.2 Smart controller with software for communication interface;
2.3 4G communication modem;
3. Industrial BUS;
4. Industrial ETHERNET point;
5. WAN INTERNET CLOUD access;
6. 802,1 lb/g router
7. Control Centre
7.2 PC Display
7.3 PC Remote monitoring software

Fig. 4. The 3D cyber-mix-mecatronic system ultraprecise multiaplicative for remote control and remote monitoring

According to Fig. 4, the matrix structure physical (mixmechatronics) and cybernetical (IT&C) of the system, enables automation, computerization and remote communication, intelligent control and monitoring, thereby contributing to raising the quality level and reducing operating costs specific to the automotive industry.

The 3D axis system (x, y, z) 1.1 with ultraprecise remote control is ordered by PC with specific software 1.4 and 1.5 for the realization of measuring function of a part 1.3 with 3D ultraprecise probe 1.2. The system is protected with laser protection barrier 1.6 and communicate with special equipments from the electronic unit 1.7 in PLC 2.1 and Internet GPRS 4G. This communication connection is linked with remote control center 3 provided with a computing station 3.2 and 3.3 connected to the router 3.1 and on which specialized software 3.4 runs.

In the 3D mode of travel, the cyber-mix-mechatronic system is designed to be operated locally using a program preinstalled on PC equipped with display and control software and modeling and emulating remote position. Switching between the two operating modes can be done anytime and measurement (3D control) points may be stored in the memory functioning in the automatic PLC mode.

Thus, all these complex functions may be implemented by integrating several functional testing and smart subsystems.

At the same time, this cyber-mixmecatronic system does not require the continued presence of the qualified and costly human operator and aims to ensure operating parameters at nominal values, along with the advantages of cyber technology, such as:

- minimizing operating – troubleshooting time;
- preventive strategies in the operation and maintenance process;
- modularity, flexibility and security.

The 3D multiaplicative cyber-mixmechatronic system, for remote control and monitoring provides connection of remote process stations to one or more central control systems, using various public or private networks for event-driven take-backs caused by an event or cyclical data processing performed using special protocols and effectively managed. The cyber-mixmechatronic system uses one or more software to connect it to the "remote control and monitoring center" based on modern GPRS technologies and multiple PLCs. The cyber-mixmecatronic system can also do the teleservice that offers data exchange via the telephone line or via the Internet and Intranet, using remote equipment and systems such as computers, machines, installations and lines production, error detection, diagnostics, activity optimization, maintenance, repairs, etc.

Thus, the cyber-mixmecatronic system makes significant contributions to minimizing the cost and increasing the efficiency and productivity of industrial activities.

The main system services consist of:

- remote control - remote control and monitoring of a system;
- remote maintenance, consist of:
  - remote monitoring – determining of the status of the system;
  - remote diagnosis – identifying the causes of malfunction;
  - remote maintenance – eliminating causes;
  - acquisition of geometric and mechanical parameters.
- **The cyber-mixmechatronic system with remote control and remote monitoring** (Fig. 5), [1, 3, 6, 7].

According to Fig. 5, the physical (mechatronics) and cybernetics (IT&C) matrix structure of the system, enables automation and computerization of remote processes (telecontrol and telemonitoring) of the industrial manufacturing line.

The system performs the function of protecting workstations from human errors specific to fabrication lines (1.1) in the series in the industry, such as automotive parts in the manufacturing industry. Parts (1.2) call feature tags with unique ID RFID (1.3) communicating bi-directionally with a drive (2.4) with automation equipment and telecommunication local but also a smart bracelet (2.3) situated on the hand of the operator who is using the device (2.2) equipped with RFID tag and bidirectional communication. Intercommunication between elements listed (part, device, operator) is collected and transmitted using antennas (2.5) and (4.1) through the Internet to a (4.2) computing station at the center of remote monitoring and remote control (4).

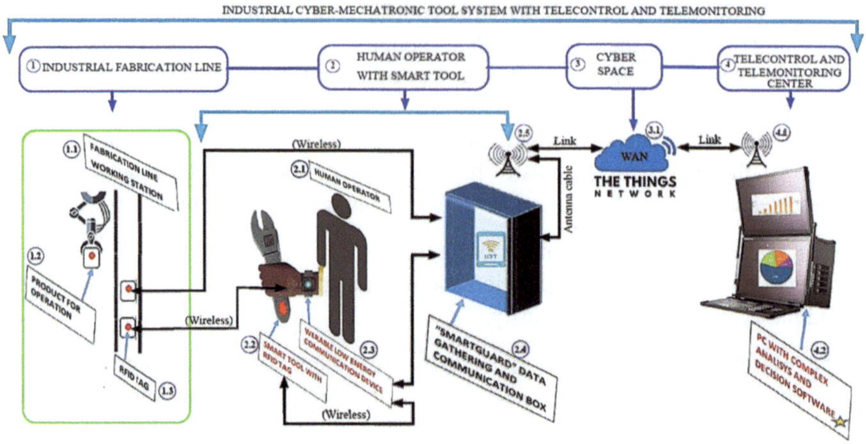

Legend:

1. Industrial manufacturing line
1.1 Workstation on the fabrication line
1.2 Product for technological applications
1.3 Unique identification system with RFID

2. Human operator with smart instruments
2.1 Human operator
2.2 Unique identification system with RFID
2.3 Smart communication and warning bracelet
2.4 Unit with system for data acquisition and communication
2.5 Antenna

3. Cyber space
3.1 Internet WAN

4. Remote control and remote monitoring centre
4.1 Antenna
4.2 PC with software for complex analysis and decision

Fig. 5. The cyber-mixmechatronic system with remote control and remote monitoring

The computer center remote monitoring and remote control (4.2) running special software designed to synchronize tasks on a database technology and eliminate errors caused by real-time tracking and manufacturing through a comprehensive analysis and forecasts.

- **Smart multi-application cyber-mixmechatronic device type industrial robot for remote control and remote monitoring of operational and service processes** (Fig. 6) [1, 5].

According to the mentioned figure, the structure of the physical matrix [Mechatronics and Cybernetics (IT & C)] of the system allows the cybernetization and remote communication of technological operational processes and of processes that service the industry, thus contributing to a higher increase in productivity and of quality of smart industrial.

The cyber-mixmechatronic multi-application system performs a remote control and remote monitoring of an industrial robot (1) connected to the cyberspace via the control unit 1.2 and the interface 1.3 with both the internal industrial bus (2.1) and Internet via a 4G GPRS modem.

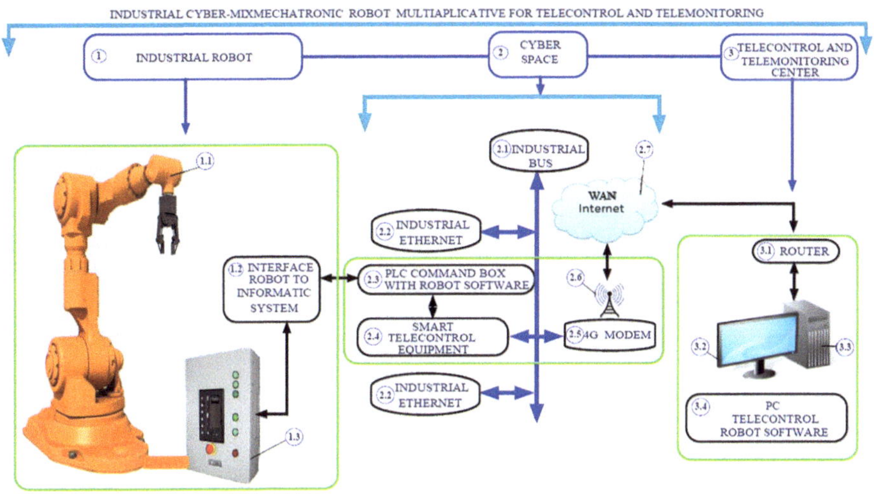

Legend:

1. Industrial robot
1.1 Universal robotic arm (with gripper / 3D feeler)
1.2 Interface between the robot and the informatic environment
1.3 Command tools

2. Cyber space
2.1 Industrial communication bus
2.2 Industrial ETHERNET network
2.3 Programmable PLC with ROBOT software
2.4 Smart devices for remote control and remote communication
2.5 4D GPRS modem
2.6 Internet WAN

3. Center of remote control and remote monitoring
3.1 Router linking to WAN Internet
3.2 PC monitor
3.3 Central unit
3.4 UPC with ROBOT remote control software

Fig. 6. Smart multi-application cyber-mixmechatronic device type industrial robot for remote control and remote monitoring of operational and service processes

Through this communication connection is made the link to a centre of remote control and remote monitoring (3) provided with a computing station (3.2) and (3.3) connected to the router (3.1) and running specialized software for robot control (3.4).

- **Cyber-mixmechatronic system for dampening for automotives and with remote control and remote monitoring functions** (Fig. 7) [1, 3].

According to the mentioned figure, the structure of the physical matrix (damper and cybernetics (IT & C) of the cyber-mixmechatronic system for dampening for automotives allows the smart computerization and cybernetization of the automotive and of the automotive industry, by raising the quality and increasing productivity of the automotive industry.

The cyber-mixmechatronic system for dampening for automotives allows the function of remote monitoring and remote control of a smart damper (1) provided with an

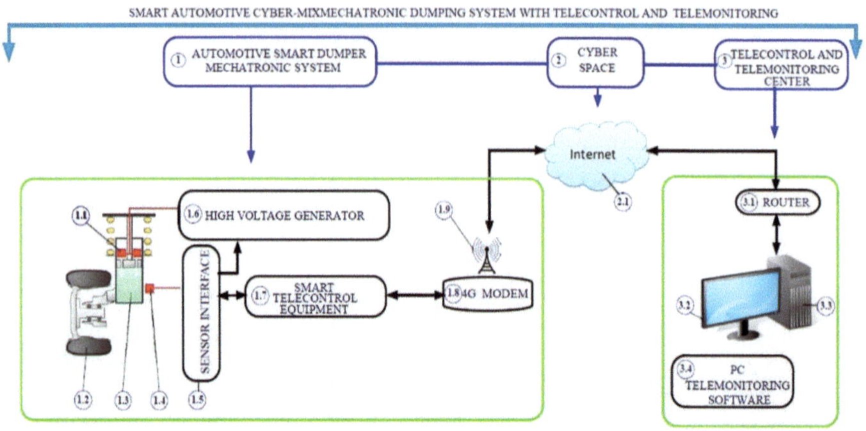

Legend:

1. Smart cyber mix-mechatronic system for dampening
1.1 Electromagnet or coil
1.2 Section of a tire
1.3 Rheological fluid
1.4 Acceleration sensor
1.5 Interface between sensor and acceleration
1.6 High tension source
1.7 Smart remote control device
1.8 4D GPRS modem
Antenna

2. Cyber space
2.1 Internet WAN

3. Centre of remote control and remote monitoring
3.1 Router linking with WAN Internet
3.2 PC monitor
3.3 Central PC unit
3.4 PC with software for remote monitoring and remote control smart DAMPER

Fig. 7. Cyber-mixmechatronic system for dampening for automotives and with remote control and remote monitoring functions

electromagnet (1.1) powered by a high voltage source (1.6) in order to modulate the degree of viscosity of the rheologic fluid (1.3).

To obtain information on the global acceleration of the cyber-mechatronic assembly attached to a tire of a vehicle uses the sensor (1.4) and the interface (1.5) connected to an intelligent control and remote control equipment (1.7).

Internet WAN network connection is performed using a specialized 4G GPRS modem (1.9) provided with an antenna (1.8).

Through this communication connection is made the link to a centre of remote control and remote monitoring (3) provided with a computing station (3.2) and (3.3) connected to the router (3.1) and running specialized software for robot control (3.4).

- **The conception and physical realization of the experimental model: "mecatronic and adaptronic & cyber-mecatronic & cyber-adaptronic system",** (Fig. 8), [1, 2, 6].

At the basis of the execution of the experimental model "mechatronic and adaptronic & cyber-mechatronic & cyber-adaptronic system" was the original conception of the logic scheme - mechatronic and adaptronic & cyber-mechatronic & cyber-adaptronic

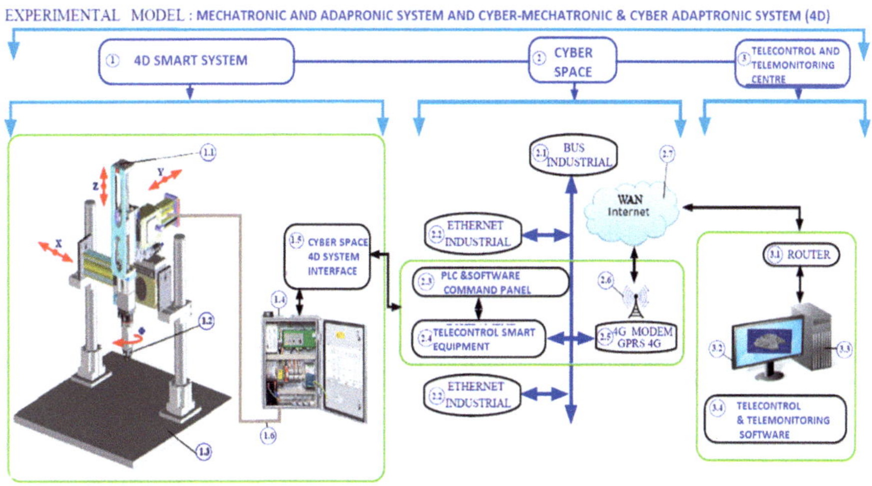

Legend:

1. **4D Smart system**
   - 1.1 4D System
   - 1.2 Ultraprecise smart gripper
   - 1.3 Anti-vibrations table
   - 1.4 Module of command and control
   - 1.5 System –cyber space interface
   - 1.6 4D System cable signals- module of command and control
2. **Cyber space**
   - 2.1 Industrial communication bus
   - 2.2 ETHERNET industrial network
   - 2.3 Command panel with PLC and software 4D system
   - 2.4 Intelligent equipment for telecontrol and telecommunication
   - 2.5 GPRS 4G Modem
   - 2.6 Antenna
   - 2.7 WAN Internet
3. **Center for telecontrol and telemonitoring**
   - 3.1 Router link to WAN Internet
   - 3.2 PC Monitor
   - 3.3 Central Unit
   - 3.4 Software for telecontrol and telemonitoring

Fig. 8. Logic scheme - mecatronic & cyber-mecatronic system 4D

system (Fig. 8) and the orthogonal projection of the experimental model with the component structure of the X, Y, Z and Φ subsystems.

Following is presented the experimental model “Mechatronic & cyber-mechatronic system in 4D coordinates (Fig. 9)”.

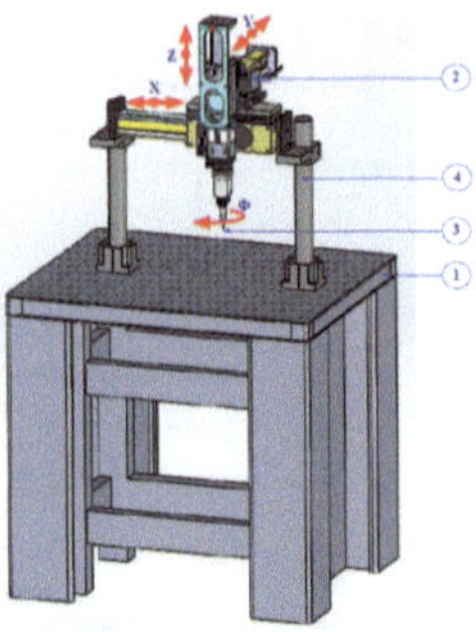

Legend:
1 - Anti-vibration table
2 – Smart system (in 4D)
3 – Smart (ultraprecise) 3D probe
4 – Adjustable support

Fig. 9. Mechatronic & cyber-mechatronic system in 4D

Following is presented the experimental model – "Mechatronic and adaptronic & cyber-mecatronic and cyber-adaptronic system in 4D coordinates" below (Fig. 10):

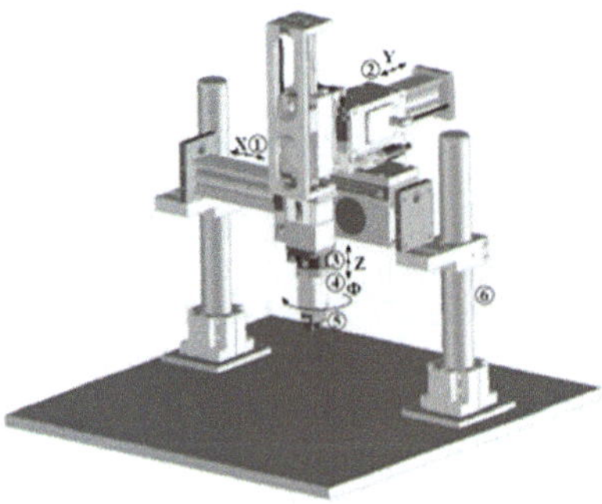

Legend:
1 – X electrical axis
2 – Y electrical axis
3 – Z – electrical axis
4 – Φ electrical axis
5 – Gripper
6 – Adjustable support

Fig. 10. Experimental *model – "Mechatronic & cyber-mechatronic* system in 4D coordinates" (without the anti-vibration table and the smart 3D probe)

According to this figure, the experimental model made consists of:

- Horizontal electric axis X: carried out in a structure defined by the nominal stroke (measuring range) of 200 mm and the gauge of the axis 510 mm by the nominal force of 205 N (and the maximum permissible force of 500 N), by 1 linear motor, 1 absolute encoded transducer with high precision and repeatability, 1 controller;
- Horizontal electric axis Y: carried out in a structure defined by the nominal stroke (measuring range) of 100 mm and a gauge length of 310 mm, by a nominal force of 120 N (and the maximum permissible force of 250 N) by 1 linear motor, 1 absolute encoded transducer with high precision and repeatability, 1 controller;
- vertical Z axis: made in a structure defined by the nominal stroke (measuring range) of 100 mm, a nominal force of 115 N (max.), 1 linear motor, 1 brake, 1 encoded absolute encoder, with high precision and repeatability, 1 controller;
- the rotary (miniaturized) electric axis and 1 controller by: torque: 0.4/1.2 (Nm); maximum rotation speed: 600 (1/min.); moment of inertia: 500 (Kgmm^2); repeatability accuracy: 0.01 (°); operating pressure: 6 (bar); meat: 1.2 (Kg); temperature:

10/40 (°C); circuit voltage: 530 (V); power supply: 1 (A); electronic control: external; encoder system: absolute encoder.

In the tests of the component subsystems and the mechatronic & cyber-mechatronics systems, the following were used:

(a) components, subsystem and smart 4G system testing program, which includes:
(a1) methods of measurement and verification of metrological parameters present;
(a2) smart mechatronic metrological metrology systems and equipment equipped with accredited RENAR (at national level) LAMs (MEMS & NEMS):

- OPTIMAHR - Germany, for measuring and checking dimensions, with nanometric precision (≤1 nm);
- 3D precision measuring machine (0.1 μm);
- Smart system with incremental rotating photoelectric transducer and digital block with digital display, in high angle and length units (≤±1 s);
- Tensiometric marker system with digital display for high precision forces and moments (Class 0.1);
- High-precision digital type (≤0.01 mm);
- Analytical balance system with high precision (≤±0.02 g);
- etc.

(a3) software programs for dimensional, angular measurements, etc.;
(a4) software programs for data analysis and interpretation (Siemens SIMATIC Panel PC 577 computing system, specialized software, etc.).
(b) documents in international catalogues relating to smart mechatronic systems and equipment for comparison and definition of metrological parameters (measuring range, precision, repeatability, fidelity, etc.);
(c) metrological documents such as histograms, diagrams, graphs and value matrices. For parametric and metrological tests, metrological standards corresponding to the equipment of the laboratories, ISO EN SR standards related to the measurement, verification and control methodologies and procedures were used.

3 Scientific Results

The scientific results obtained in the paper are synthesized as follows:

- **concepts of virtual models** in variants and scenarios for mechatronic and cyber-mechatronic systems, with applications in industry (e.g. automobile industry - SC Automobile Renault - Dacia SA Pitesti, SC Componente Auto SA Topoloveni, SC Comis SRL Valenii de munte etc.) in medicine (e.g. smart mechatronic technologies for selective sintering with laser beam - Hospitals in Bucharest, etc.);
- Physical models in original solutions for mechatronic and cyber-mixmechatronics systems with industrial applications (e.g. the fine mechanics and mechatronics industry - Mechatronics and Cyber - MixMechatronics Laboratory at INCDMTM Bucharest, for measurements, micro and nano technologies for Ph.D. thesis, from the

Doctoral School of Mechanical Engineering and Mechatronics of UPB and UVT, etc.);
- real parametric values for remote metered and mechatronic systems: telecontrol, telemonitoring, teleservice and teleconfiguration;
- process validation and mechatronic and cyber-mix-mecatronic system in real industrial applications.

4 Conclusions

By carrying out this scientific work "Systems and Intelligent mechatronics and cyber-mixmechatronics ecosystems developed in "ECOSIN-MECATRON", research infrastructure", the author presented:

- the creation and development of a new, specialized 21st century intelligent specialized domain to ensure the sustainable development of the national and international economy and industry, the field of Mechatronics and Cyber-Mix-Mechatronics;
- creation and development of new mechatronic and cyber-mixmechronic systems, in original and multi-purpose applications in many industrial, economic and societal environments;
- substantial contributions to solving the scientific paradigm Mechatronics, MixMecatronics and Cyber-MixMecatronics;
- Original contributions to research, development and innovation in the intelligent specialized fields of Mechatronics, MixMecatronics (Integral and Adaptronics) and Cyber-MixMecatronics.

References

1. Gheorghe, Gh.: Mechatronics & Cyber-Mechatronics Systems. CEFIN, Bucureşti, România (2015)
2. Gheorghe, Gh.: Smart Adaptronic Micronanoengineering. CEFIN, Bucureşti, România (2014)
3. Gheorghe, G., et al.: Mechatronics and micro – mechatronics concepts and techniques for intelligent integrating micro engineering. In: The 19th International DAAAM Symposium "Intelligent Manufacturing & Automation: Focus on Focus on Next Generation of Intelligent Systems and Solutions", 22–25 October 2008
4. Giurgiutiu, V., Lyshevski, S.E.: Micromechatronics: Modeling, Analysis and Design with MATLAB, p. 856. CRC Press, ISBN: 084931593X
5. Jănker, P., Hermle, F., Lorkowski, T., Storm, S., Wettemann, M., Gerle, M.: Actuator technology based on smart materials for adaptive systems in aerospace. In: Proceedings of ICAS 2000, Harrogate, UK (2000)
6. Gheorghe, G.: Study and analise regarding the concepts of mechatronics and adatronics virtual models and cyber-mechatronics and cyber-adaptronics. Cefin Publishing House, Romania (2016)
7. Gheorghe, G.: Elaboration of new scientific and technological concepts and new virtual mechatronics and adaptronics models and cyber-mechatronics and cyber-adaptronics. Cefin Publishing House, Romania (2016)

8. Bishop, R.H., Erbe, H.: The Mechatronics Handbook. The University of Texas at Austin, Austin (2002)
9. Bruns, F.W., Erbe, H.: Engineering future laboratories. In: Marvel – A Leonardo da Vinci Pilot Project– Mechatronics Training in Real and Virtual Environments, Bremen (2005)
10. Pereira, C.E.G., Paladini, F.: Control and automation engineering education: combining physical, remote and virtual labs. In: Proceedings of the 9th International Multi-Conference on Systems, Signals and Devices – SSD 2012, Chemnits, Germany, pp. 1–10, 20–23 March 2012
11. Pop, L.G., Matiec, V.: Transdisciplinary approach of the mechatronics in the knowledge-based society. In: Martinez-Alfaro, H. (ed.) Advances in Mechatronics. InTech (2011)

Author Index

G. I. Gheorghe (Ed.): ICOMECYME 2018, LNNS 48, pp. 225–226, 2019.
https://doi.org/10.1007/978-3-319-96358-7

MIX
Papier aus verantwortungsvollen Quellen
Paper from responsible sources
FSC® C105338

If you have any concerns about our products,
you can contact us on
ProductSafety@springernature.com

In case Publisher is established outside the EU,
the EU authorized representative is:
Springer Nature Customer Service Center GmbH
Europaplatz 3, 69115 Heidelberg, Germany

Printed by Libri Plureos GmbH
in Hamburg, Germany